Energy, Climate and the Environment

The aim of this series is to provide texts which lay out the technical, environmental and political issues relating to proposed policies for responding to climate change. The focus is not primarily on the science of climate change, or on the technological detail, although there will be accounts of this, to aid assessment of the viability of various options. However, the main focus is the policy conflicts over which strategy to pursue. The series adopts a critical approach and attempts to identify flaws in emerging policies, propositions and assertions. In particular, it seeks to illuminate counter-intuitive assessments, conclusions and new perspectives. The intention is not simply to map the debates, but to explore their structure, their underlying assumptions and their limitations. The books in this series are incisive and authoritative sources of critical analysis and commentary, clearly indicating the divergent views that have emerged whilst also identifying the shortcomings of such views. The series does not simply provide an overview, but also offers policy prescriptions.

More information about this series at
http://www.palgrave.com/gp/series/14966

Theresia Betty Sumarno

Just Energy Transitions and Coal Bed Methane

The case of Indonesia

palgrave
macmillan

Theresia Betty Sumarno
CEPMLP
University of Dundee
Dundee, UK

Energy, Climate and the Environment
ISBN 978-3-030-85489-8 ISBN 978-3-030-85490-4 (eBook)
https://doi.org/10.1007/978-3-030-85490-4

Cover illustration: RelaxFoto.de

This Palgrave Macmillan imprint is published by the registered company Springer Nature Switzerland AG.
The registered company address is: Gewerbestrasse 11, 6330 Cham, Switzerland

This book is dedicated to my ***Uncle Priest****,* ***Romo Isidorus Warnabinarja, SJ (RIP)*** *for his blessings and prayers from heaven as well as my parents, Bapak* ***Joseph Sumarno*** *and Ibu* ***Maria Goretti Supeni****; my brothers, Ignatius Agung Sumarno (****Mas Agung****) and Paulus Cahyo Sumarno (****Mas Paul****); my sisters-in-law, Sara Kirmiyati (****Mbak Ate'****) and Dewi Ketaren (****Mbak Dewi****); finally, my favourite nephews, Samuel Ekaristi Daniswara (****Mas Muel****), Nikolaus De Paul Vitonata (****Bang Niko****), and Theodorus Delano Giannesta (****Adek Theo****).*

Preface

I was born in DI Yogyakarta, a special district in Indonesia, located in Java Island. "What do you want to be when you grow up?" This question is what mostly people ask us when we were little. I was always dreaming to study abroad and learn different things in life. Happened to be born in a lower-middle-range family, I had no clue how I would get that privilege to be having experiences overseas. Most of funding opportunities that were available at that time were for postgraduate studies and mainly for civil employees.

Four years doing a PhD in the UK in energy economics and policy management area has opened different opportunities in my career. I never thought to be one of the energy experts who works both nationally and internationally. Energy excites me, and being an Indonesian made me realise that energy is one of the biggest challenges in my country. I have learnt how global world respond to the climate change. Indonesia is one of the largest archipelago countries, where more than 270 million people live on different islands with different levels of energy supply and demand. On top of that, Indonesia has abundant energy resources, both fossil fuel and renewable energy sources. This fact has fascinated me and driven me to write on how diversifying energy resources could help Indonesia to transition to a cleaner energy.

Energy has played important role in countries' economy, including Indonesia. Blessed with its geographical location and natural resources have made Indonesia to be one of the most resourceful in ASEAN. It is like two sides of knife where geographical situation also creates more challenges to Indonesia when it comes to developing every island. There is an extra need of developing adequate infrastructures, including its energy infrastructures. Indonesian population has been increasing from time to time, which therefore increases Indonesian energy demand. It becomes more challenging for Indonesia to meet its energy demand while needs to meet its Nationally Determined Contributions (NDC) targets.

This book provides different perspectives for Indonesian energy sector based on literature, expertise interviews, focus group discussion, and field survey in local communities. The topic being chosen here is CBM development in Indonesia in the era of low-carbon transition in Indonesia. Indonesia has abundant resources of CBM and had been through different challenges since it was first initiated. Being part of new energy sources stipulated in the national energy policy, CBM could support the Government of Indonesia to reach its NDC.

Dundee, UK Theresia Betty Sumarno

Acknowledgement

I thank the Lord for has made things impossible, possible. I have always wanted to write a book, and this is my first book. This book was written during the COVID-19 pandemic, and I would like to thank those who played important roles in the process.

Thanks to Dr Geoffrey Wood and Palgrave Macmillan for allowing me to contribute to the world of knowledge about this specific topic. I also would like to thank those who never stopped supporting me, Mr Sudirman Said, Dr Widhyawan Prawiraatmadja, Dr Andang Bachtiar, and all the National Exploration Committee, Republic of Indonesia, and in particular Government of Indonesia, especially for the LPDP Scholarship that played a significant role in my journey. To those who have never lost faith in me, Dr Mohammad Barkindo, Rini Saputra, Dr Parulian Sihotang, and all of my friends who live in different places.

This book required comprehensive data so I would like to take this opportunity to thank all the participants in this book: the interviewees (investors, government, and consultancies), the members of the focus group discussion and support from the Association of Oil and Gas Producer Areas, the local communities, and the students who helped me during the survey (Balikpapan, Banjarmasin, Palembang, and Riau).

Some of the parts of this book were reviewed by experts in their particular fields prior to submission; therefore, I would like to thank Professor Diane Kraal, Daniel Johnston and David Johnston, Professor Bruce Burton, Dr Ariel Bergmann (RIP), Stephen Dow, and Professor Raphael Heffron.

Last but not least, to my dear family members (Bapak, Ibu, Mas Agung, Mas Paul, Mbak Ate, Mbak Dewi, Samuel, Niko, and Theo), who are away in Indonesia and who never stop sending their prayers for me at all times and funny videos to keep me smiling.

Finally, I hope that this book inspires not only those who read it but also my loved ones, especially my three most loved nephews.

Contents

Abbreviations

ADPM	*Asosiasi Daerah Penghasil Migas* (Association of Oil and Gas Producing Areas)
ASEAN	Association of South East Asian Nations
ASR	Abandonment and Site Restoration
Bboe	Billion barrels oil equivalent
BCF	Billion cubic feet
Bcm	Billion cubic metre
CBM	Coal bed methane
CeCon	Coal bed methane economics
CIT	Corporate income tax
CMoEA	Coordinating Ministry of Economic Affairs
CSR	Corporate social responsibility
DCF	Discounted cash flow
Ditjen Migas	Directorate General of Oil and Gas Republic of Indonesia
DMO	Domestic market obligation
DSP	Drilling site preparation
EBA	Environmental basement assessment
EIA	Energy information administration
EOR	Enhance oil recovery
ERR	Effective royalty rate
FDP	Field development plan
FGD	Forum group discussion
ft	feet

FTP	First Tranche Petroleum
G&G	Geology and geophysics
G20	The group of twenty
GCA	Gas cost allowance
GDA	Geosains Delta Andalan (Geology Consulting Firm)
GoI	Government of Indonesia
GR	Government Regulation
HSE	Health, safety, and environment
ICP	Indonesian crude price
IISD	International Institute for Sustainable Development
IMF	International Monetary Fund
IPA	Indonesian Petroleum Association
IRR	Internal rate of return
LNG	Liquefied natural gas
LTBR	Long-term bond rate
m	Metre
m3/t	Metre cubic per ton
Mbtu	Thousands British thermal unit
mD	Millidarcy
mmbbl	Million barrels
mmscfd	Million standard cubic feet per day
MoEF	The Ministry of Environment and Forestry
MoEMR	Ministry of Energy and Mineral Resources
MoF	Ministry of Finance
MoHA	Ministry of Home Affairs
MoM	Ministry of Manpower
MoMAF	The Ministry of Marine Affairs and Fisheries
MPa	Mega Pascal
MR	Ministry Regulation
NCF	Net cash flow
NCR Gross PSC SS	Non-Cost Recovery Gross Production Sharing Contract Sliding Scale
NEC	National Exploration Committee
Net PSC SS	Net Production Sharing Contract Sliding Scale
NOC	National Oil Company
NOECD	Non-Organisation for Economic Cooperation and Development
NPV	Net present value
OECD	Organisation for Economic Co-operation and Development

PCOS	Producer cost of service allowances
PHE TJ	Pertamina Hulu Energy Tanjung
PHE	Pertamina Hulu Energy
POD	Plan on development
POS	Point of sales
POV	Point of value
PR	President Regulation
PRB	Powder River Basin
PRRT	Petroleum resource rent tax
PSC	Production sharing contract
PTK	Guidelines and working procedures
PWC	PriceWaterhouseCoopers
R&D	Research and development
R/T	Royalty tax
RE	Renewable energy
ROR	Rate of return
Rp	Rupiah
RRR	Reserve replacement ratio
RUEN	*Rencana Umum Energy Nasional* (General Energy National Plan)
SKKMigas	Satuan Kerja Khusus Pelaksana Kegiatan Usaha Hulu Minyak Dan Gas Bumi (Special Task Force for Upstream Oil and Gas Business Activities Republic of Indonesia)
SP	Selected price
SPE	Society of Petroleum Engineers
SS	Sliding Scale
TcF	Trillion cubic feet
UCG	Underground coal gasification
WP&B	Work, programme, and budget

List of Figures

List of Tables

1

Introduction

Abstract This chapter provides an overview and context of the book itself, that is, an overview of Indonesia, energy transition, energy justice, CBM, and taxation, by discussing the aims of the book in terms of the energy transition movement particularly for developing and emerging economies.

Keywords Australia • Coal bed methane • Fiscal regime • Fiscal terms • Indonesia • Petroleum Resource Rent Tax (PRRT) • Production Sharing Contract (PSC) • Ring fence • Royalty/tax • Wyoming • The US

1.1 Introduction

This chapter provides an overview and the background context of the book. The focus is on the energy transition in Indonesia from an energy justice perspective and exploring the contribution of coal bed methane (CBM) to Indonesia's National Energy Policy (NEP) and the Indonesian energy mix. Also explored are the incentivisation schemes through

T. B. Sumarno, *Just Energy Transitions and Coal Bed Methane*, Energy, Climate and the Environment, https://doi.org/10.1007/978-3-030-85490-4_1

taxation policy that are applied to develop energy technology (in terms of CBM) in Indonesia, the US and Australia. A recurrent theme throughout the book is that within the transition process to a low-carbon economy, it is important to consider that justice is present where the benefit and impacts of the energy sector are distributed equally to society.

As one of the most populated countries in South East Asia (SEA), Indonesia accounts for around 40% of the total population in SEA, and hence the development of a cleaner energy movement in Indonesia becomes really significant. Indonesia is the world's fourth most populous country with a current population of 274 million people in 2020 (United Nations 2020b), and is forecasted to increase to 331 million people in 2050 (Indonesia Population, Worldometer 2020). This increase in population and growth in the economy will increase energy demand in Indonesia. Indonesia is now classified as an upper-middle-income country (World Bank 2020a) and has the biggest economy in SEA (World Bank 2020b). Indonesia is also the world's biggest island country consisting of more than 17,000 islands, around 6000 of which are inhabited. Any success in the country's energy development plan will contribute to the rest of the region and also could act as a guide for neighbouring countries and other countries with similar contexts and challenges.

This book gives a new perspective on the energy sector, especially on CBM and its role in the energy transition in Indonesia. Further, this book will also add to the literature on the energy transition as seen from the role of the fossil fuel perspective, which in this case is CBM. By bringing new perspectives to advance the energy transition in Indonesia together with other energy sources, it is hoped that this book will lead to energy sustainability, providing readers with an understanding of the key issues around CBM project development in general and its role in a just transition to the low-carbon economy in Indonesia and beyond. For resource-rich (in particular for CBM) developing and emerging countries that are moving towards the low-carbon economy, this book also seeks to provide readers with food for thought for their transition strategy.

1.2 Overview on Indonesia's Energy Sector

According to the National Energy Council of the Republic of Indonesia (2020), Indonesian's energy demand is going to be dominated by three different sectors until 2050: industry, transportation, and households. Industry and households mainly use coal and gas as their energy sources, while transportation uses fossil fuel. According to Indonesian Presidential Regulation No.22/2017 (General Plan of National Energy), industry is estimated to use 48% of total final energy, while transportation and household are estimated to use 30% and 15%, respectively, of total final energy by 2025. Assuming economic growth is 7.3% in 2050, these three sectors will still be the main users of the total energy being produced with lower percentages compared to the 2025 energy demand mix, which are 46%, 26%, and 14% for industry, transportation, and household, respectively.

That being said, fossil fuels remain the main energy sources in 2025: oil (25%), gas (22%), and coal (30%). However, Indonesia has a general plan for renewable energy (RE) development, with a target of a 23% share of RE in the energy mix by 2025. This energy mix will evolve and is estimated to change by increasing the usage of RE to 31% in 2050, and reducing the usage of fossil fuels, especially coal to 25% in 2050 and oil to 20%, while gas usage is estimated to increase to 24%. Increasing RE usage and decreasing fossil fuel usage are expected to increase greenhouse gas (GHG) emission reduction from 34.8% in 2025 to 58.3% in 2050. A review of the NEP achievement, published during the fourth quarter of 2020, and the NRE 23% in 2025 is far from the current achievement. Indonesia has only achieved 9.5% NRE since 2015, meaning that in five years' time Indonesia must add more to the equation, that is, 13.5% RE to its energy mix (National Energy Council 2020). CBM is known as a new energy technology in Indonesia; therefore, it can contribute to reducing carbon dioxide emissions through reducing a reliance on coal in Indonesia.

According to the NEP (Government Regulation No. 79/2014), CBM was envisaged to play a very minor role in the national energy mix by 2025. Bachtiar (2018) has noted that the Government of Indonesia (GoI) does not seem to be intentionally developing the CBM industry regardless of their effort all the way back in 2010 when they first tried attracting investors to develop CBM in the country.

Indonesia's CBM resources are located in the archipelago basins with total potential CBM resources of 453 Tcf (Wood Mackenzie 2015; CBM Asia 2016; Stevens et al. 2001; Stevens and Hadiyanto 2004). Of these basins, 85% contain lignite to sub-bituminous (low-rank coal) and 15% is bituminous to anthracite (low-rank coal) (Flores 2014a: 568). CBM operators initially believed that Indonesian CBM was not a prospective resource due to its rank and low quality until the success of the Powder River Basin (PRB), the largest coal bed basin in the US (Stevens and Hadiyanto 2004). The sub-bituminous coal has similarities with the PRB in terms of coal thickness, coal rank, areal extent, and gas adsorption capacity (Flores 2014b). Moreover, the Miocene coal deposits in Indonesia have a thicker, deeper, and higher rank of coal than the PRB does (Stevens and Hadiyanto 2004), and therefore Indonesia is expected to have far better resources.

The first seven CBM projects were awarded in 2008 under Production Sharing Contract (PSC) arrangements. This CBM PSC applies the same terms and regulation with PSC for natural gas, the only difference being the split of its CBM, which is 55:45, whereas for natural gas the split is 70:30, for government and investor, respectively. The number of CBM projects increased to 54 contracts in 2012, but these numbers decreased by 2018 to just 30 contracts (SKKMigas 2018) with no CBM production. This highlighted the point that there is more to a successful energy sector than just geological conditions. What also matters is the taxation system which plays a significant role in any energy resource development (Collier 2010).

1.3 The Non-OECD Vs OECD Countries' Fiscal Regimes and the Importance of Fiscal Incentives

According to Collier (2010), for low-income resource-rich countries (such as Indonesia), resources discovery is likely to be far more critical than in high-income resource-rich countries. In some low-income resource-rich countries, there are bountiful resources that have not been discovered. The reason is in part due to country-specific fiscal regime arrangements that do not provide sufficiently attractive incentives to discover the resources in the first place (Collier 2010). Vivoda (2017) has argued that resource-rich countries with high-grade deposits attract investment more easily. From the investors' perspective, those who are risk-taking juniors will aim to invest in areas that have good prospects regardless of risks. However, when it comes to investment decision-making, they will balance the prospects against risk criteria (Vivoda 2017). According to some, geology is the most influential factor that determines countries' investment attractiveness, while regulatory and fiscal are the fourth and fifth most influential factors (Johnson 1990; Otto 1992; O'Neill 1993; Morisset 1999; Morgan 2002; Kasatuka and Minnitt 2006; Tienhaara 2006; Penney et al. 2007; Tole and Koop 2010). However, Collier (2010) mentioned that in order to manage the resources successfully, there is a second aspect considered equally as important as the discovery of the resources, which is the design of appropriate fiscal regimes. The quality of the hydrocarbons, the amount discovered, and the location of the reservoir and its proximity to consumer markets are known as the main factors that determine a commercial discovery (Sigam and Garcia 2012).

Many developing economies use tax incentives to attract both domestic and foreign investment, as for them the revenue generated from natural resources are often their primary source of revenue (Redhead 2018; IMF 2015). The extractive industries are significant for many developing countries to grow their economy (United Nations 2017; Sigam and Garcia 2012). For many resource-rich developing countries, revenues generated from extractive industries are often the major sources of revenue (Redhead 2018). The government should create a balance or sustain a supportive environment for substantial investment with the country's needs, such as revenue that can be used to support their development (United Nations 2017). Some developing economies may be more concerned to show that the one who exploits resources should be the nation rather than foreign companies (Calder 2014). Meanwhile, in the OECD countries, it is common to have privately owned companies carry out natural resource operations (Calder 2014). However, Kasatuka and Minnitt (2006) found that there were some non-commercial risks that deterred foreign investment in resource-rich countries, including government instability, poor socio-economic conditions, and corruption.

In contrast to developed economies, significant revenues generated from extractive industries in developing economies can often lead to poor governance. Therefore, the implementation of a contractual agreement and state participation is significant (Calder 2014). Calder (2014) also wrote that many developing economies imposed corporate income tax as well as output-based royalties, and countries may also impose additional taxes to obtain a higher share of the revenue. In consequence, countries end up creating a fiscal regime that is not easy to administer as they imposed graduated royalties to obtain higher shares yet still imposed an income tax. It has become a challenge to administer fiscal regimes applied to many different varieties of resources. Therefore, it is known that many developing economies have no one-size-fits-all fiscal regime (Calder 2014; United Nations 2017). Designing appropriate fiscal regimes which provides sufficient revenue for the country while also providing enough encouragement to invest is vital, as well as the necessary supportive legal and policy framework (Redhead 2018; Calder 2014; IMF 2012; Sigam and Garcia 2012). This becomes the main task for policymakers (Redhead 2018; Calder 2014; IMF 2012). Meanwhile, fiscal decentralisation is more common in developed economies, such as in the US and Australia with federal government structures (Calder 2014).

1.4 Discussions on CBM PSC in Indonesia, the US Royalty/Tax System, and the Australian Petroleum Resource Rent Tax

1.4.1 CBM PSC in Indonesia

There have been three generations of CBM fiscal regime since the first CBM block was awarded in 2008. The oldest generation of CBM PSC was implemented only from 2008 to 2009. The second generation was Net PSC with different terms on production split before POD, and the last generation is the Gross Split PSC, implemented since January 2017.

Indonesia has been exploring for unconventional hydrocarbon since 2008. Sukhyar and Fakhruddin (2013) stated that Indonesia has identified 128 sedimentary basins, and that most of the basins are indicated to contain unconventional hydrocarbons (shale oil and gas and CBM). Indonesia has one of the world's largest CBM resources (Len 2014). The first CBM PSC contract was signed in 2008, and up to 2016 there were 54 CBM PSCs which had been signed (PriceWaterhouseCoopers 2014; National Exploration Committee 2015; Bowe and Moore 2015). Indonesia set a regulation to improve CBM development activities, namely MoEMR Regulation No. 36/2008. This regulation was expected to attract new investors on CBM (PriceWaterhouseCoopers 2016). There are two PSC generations on CBM; the first generation was those signed in between 2008 and May 2009. The difference between the first generation and the second generation was the treatment for production before POD and the cost recovery. The first CBM PSC had 90% cost recovery, and the GoI owned the production before the POD. In the second CBM PSC, the cost recovery was 100%, and the contractor could sell the production before POD and share it based on the split in the contract. During this period, the cost was not yet recovered (PriceWaterhouseCoopers 2016).

The MoEMR established a regulation for CBM, Ministry Regulation No. 36/ 2008. This regulated CBM activities, covering the licensing process and sharing facilities in the activities. Chapter 8 article 31 of the regulation stipulated that CBM was one of Indonesia's potential strategic resources to fulfil domestic needs. An optimised development would, therefore, be required for CBM. The CBM contract could be publicly tendered or a direct offer by the joint study or joint evaluation.

In late 2015, the MoEMR published a new regulation for CBM acceleration, Ministry Regulation No. 38/2015, regulating the options of the fiscal system for CBM as a result of discussions between the GoI and the CBM contractors. The first option was the Net PSC. Similar to the PSC, this applied to conventional hydrocarbons with modified shares of 55:45 (GoI: CBM contractors) to attract prospective CBM investors. Adding favourable long-term CBM prospects to Indonesian (Len 2014), these modified shares attracted many CBM investors.

The current CBM contractors have faced many challenges very much related to PSC implementation. Therefore, the two other options were designed as a response to the problems and challenges faced by the contractors. The second option was the Net PSC Sliding Scale proposed by SKKMigas. The difference here was in the implementation of the Sliding Scale system on First Tranche Petroleum (FTP) and the production split after cost recovery. The third option was a non-cost recovery Gross Split PSC Sliding Scale proposed by the CBM investors.

Based on research on fiscal regimes in some CBM countries around the world, Chakhmakhchev and Fryklund (2008) found that Indonesia was predicted to meet difficulties in developing the CBM industry with its current fiscal regime. The fiscal system currently applied to CBM in Indonesia does not support the development of the industry. Favourable geological conditions are insufficient to develop the industry. Other things have to be appropriately set up to support the development of the industry, such as the fiscal system, regulation, as well as infrastructure (Sigam and Garcia 2012; IMF 2012; Redhead 2018).

Having oil and gas businesses with PSC as a system requires the players to comply with regulations attached to the PSC to allow them to recover their costs, and any unexpected costs that are not eligible for cost recovery (e.g. any costs that are not approved by the GoI as part of their activities budget) may distort the project value (Metters 2014). In the PSC implementation, investors operate the business activities at their own sole risks, and only when they start producing the hydrocarbons do they start to share their risks with the government via the cost recovery mechanism (Metters 2014). In the third-generation PSC (after 2010), investors were no longer able to transfer any cost in a dry hole field to the production field due to the ring-fence policy (Metters 2014). According to Metters (2014), cost recovery is a control mechanism by the GoI on the development of their resources which were carried by investors. Hence, cost recovery is not a right for investors, rather a result of investors complying with the regulations in conducting their business in Indonesia.

Indonesia started its first CBM project in 2008 and applied the conventional PSC to CBM. CBM development in Indonesia was expected to increase both gas reserves and increase the use of environmentally friendly sources of energy and employment (Makmun and Sitepu 2012). One of the most important aspects in developing CBM projects in Indonesia is designing an economic scheme to govern the shares between the GoI and investors (Lalean 2008). Four years later, CBM players had become fully aware that the business models applied to CBM were completely different from the business models of conventional oil and gas (Lalean 2012). CBM required more flexibility to accommodate the dynamics of the CBM operations, and the initial CBM PSC applied in Indonesia was considered unattractive (Lalean 2012; Makmun and Sitepu 2012). A CBM project in Indonesia is classified as a marginal project. Hence, a supportive regulation and fiscal system has become significant for CBM development (Makmun and Sitepu 2012).

Furthermore, the CBM business model cannot simply be adopted from conventional oil and gas models. CBM capital expenditures and operating expenditures in each CBM field can be different from one to the other (Lalean 2008). According to Lalean (2010), the government should develop a supportive policy to improve CBM investment. Kusumah and Munawar (2015) noticed that CBM development in Indonesia has been slowing down since 2015, highlighting that it would be hard to develop a greenfield project[1] even with huge resources (in the case of a CBM project) with the initial CBM PSC since a CBM project is capital intensive, especially at the initial stage (Kusumah and Munawar

[1] Greenfield project means the project does not follow any previous project; in other word, the project requires new facilities to kick-off (Bayar 2017).

2015). Not only is a huge investment fundamental for CBM to kick off, it also requires land access which is one of the challenges facing CBM development in Indonesia (Kusumah and Munawar 2015). However, if the CBM project is a brownfield project,[2] it still can be economically viable using the initial CBM PSC with conditions where there is full integration with a conventional project, and there is flexibility on its regulation that allow this to happen (Kusumah and Munawar 2015). Incentives on tax to the CBM industry in the US is one of the key factors of CBM development success there (Chakhmakhchev and Fryklund 2008). CBM projects in Indonesia require high gas prices as well as supportive fiscal mechanisms (Makmun and Sitepu 2012). Lalean (2012) stated that Indonesia must re-engineer the operating principles and procedures for the CBM industry, and the GoI must recreate a new fiscal system applied to CBM (Lalean 2012). According to Len (2014), the GoI should consider a new administration system that enables greater cost-effectiveness and project flexibility. These can be done by addressing the licensing and fiscal regimes, as well as the regulatory and bureaucratic constraints, and procurement restriction (PTK 007) in CBM activities (Len 2014). Makmun and Sitepu (2012) stated that the next step after improving the regulation would be to develop an effective fiscal system for the industry.

This book will later discuss comprehensively in the Chap. 6 the implementation of CBM PSC in Indonesia and provide more evidence of the importance of the efficient and effective fiscal system to support the CBM industry in Indonesia by looking at the US Royalty Tax System and the Australian Petroleum Resource Rent Tax System. Therefore, discussion on how the US and Australia implemented their fiscal policy, including incentives for CBM, are discussed in the following subsections.

[2] Brownfield project means the project utilises existing facilities, or sharing facilities with other project (Bayar 2017).

1.4.2 The US Royalty/Tax System Applied to the CBM Industry

The implementation of the royalty/tax system in the US is somewhat different depending on whether implemented at the state or the federal jurisdiction. For example, in terms of tax parameters, among major oil-producing states, tax structures vary considerably, and tax bases interact differently between the state and the federal level (Kunce et al. 2003). These differences can be found in the treatment of royalty payments, incentives on exemptions, and credits against tax liabilities in a specific condition (Kunce et al. 2003).

Royalty paid to the federal government as a result of oil and gas production on federal land contributes significantly to federal government revenues. However, 20% of this revenue is usually transferred to the state governments for funding their public services (Bank and Kuuskraa 2006). In the State of Wyoming, for example, state royalties and production tax income have become important sources of state revenue and public service development (Bank and Kuuskraa 2006).

One of the taxes in Wyoming State's fiscal regime for CBM is severance tax. According to the model by Kunce et al. (2003), increasing severance tax can decrease production in two ways. First, increasing severance tax will reduce the future net-of-tax profits, which in turn limits the exploration incentives and eventually reduces exploration activities. Fewer exploration activities leads to less additional reserves being discovered, which then increases exploration costs, and finally decreases the production costs. Second, if the exploration effort is constant, increasing severance tax will reduce the net profit, and, at some point, it will reach marginal production, and eventually will cease the production. They also stated that in the case of increasing the severance tax in Wyoming, this only reduced production by less than 6% over a 40-year period; hence, overall the state will increase their revenue from severance tax (Kunce et al. 2003).

CBM development (PRB) in Wyoming has grown dramatically since the late 1990s, which contributed significantly to the state government and its counties' income from taxes and royalties (Mackinnon and Fox

2006). The CBM industry has brought impacts to the economy of Wyoming. Not only has it contributed to state revenues (i.e. severance tax), it also increased the employment rate in Wyoming (Waeckerlin 2005). It was expected to create approximately 7000 new jobs in 2006. In that time, the CBM development brought improvements to their public programs, such as education and healthcare (Waeckerlin 2005). A report by Anderson ZurMuechlen and Co. (2001) revealed that there has been multiplier effects on CBM development in the PRB. CBM development also increased the revenue of the government from the perspective of employee's wages and benefits. Thus, it can be stated that new jobs, new income sources, and new revenue for the government are the results of CBM development in the PRB (Bryner 2002; Hulme 2002; Mudd 2012). These do however have multiplier impacts on society, such as the increase in demand on public services and increasing environmental impacts including traffic, air pollution, and noise (Bryner 2002; Hulme 2002).

Having such a huge contribution to both federal and state governments, the US federal government has supported and offered incentives for CBM development activities. One of the incentives given at the national level is tax credits for unconventional technologies used for extracting oil and gas. In addition, the federal government provided research support for CBM development through the Department of Energy and land access to federal land for CBM development (Skov and Myers 2004). The Wyoming legislature urged the state agencies to speed up the permit issuance process for CBM production (Skov and Myers 2004).

As mentioned above, as a multiplier effect of CBM development, it also results in environmental impacts to the state. There is a significant body of literature that has raised concerns about the resultant impacts on water quality and quantity. Fracking and dewatering are the two main activities that are most likely to impact water (Bryner 2002; Waeckerlin 2005; Hulme 2002; Skov and Myers 2004; Mackinnon and Fox 2006; Mudd 2012). This literature states that the dewatering process could lead to diminished underground water supplies, and there is a risk of groundwater being polluted when discharging the water. There may be circumstances where the environment is unable to handle a large amount of water being disposed of as a result of the dewatering process in CBM development. Bryner (2002) and Hulme (2002) added that more environmental disturbances such as noise, air pollution, increasing traffic, congestion, and issues with other land uses can also happen as a result of CBM development.

Hulme (2002) also mentioned that air pollution was one of the impacts of CBM activities directly related to health conditions experienced in CBM areas. In addition, there is also a concern for wildlife habitat as there is an increase in human contact in the occupied areas. These impacts have been mitigated by some governments (such as the State of Wyoming) since CBM has significantly contributed to economic growth and its national energy production (Mudd 2012). Likewise, the Government State of Colorado has established mitigation rules and the requirement, for example, for surface mitigation, and is also required for CBM players to install electric motors to reduce noise levels (Bryner 2002).

Therefore, while there are many positive contributions to the states, CBM development also has created additional challenges to the government. There has been significant population growth in the areas where CBM developments are located (Hulme 2002; Bryner 2002). This then becomes a challenge for the government since they have to further develop public facilities including growing demand for housing, water, sewer, education, health care services, and refuse disposal. More funds are required to be allocated to face these challenges. As a result, CBM development in these states has allowed the government to spend more fund for education (i.e. school facility improvement) (Hulme 2002).

1.4.3 Discussion on the Petroleum Resource Rent Tax System

The CBM industry in Australia (CBM is sometimes also called 'coal seam gas') is also known to have contributed to economic and regional development in Australia (i.e. increases household income, GDP, government revenue), and increased numerous employment opportunities for the relevant community (Cully 2015). As in the PRB, the Australian government has attempted to achieve a balance between maximising the benefits of the industry and managing the environmental risks (Letts 2012).

The fiscal arrangements in Australia were considered to be among the most competitive fiscal arrangements worldwide (Australian Government Geoscience Australia 2015). There had been a long process of development, and the Australian government eventually decided upon the Petroleum Resource Rent Tax (PRRT) for their petroleum industry. Theory on the economic rent of mineral resources had been known since 1914 via Lewi Gray, but the theory on taxing the economic rent was only written in 1948 by E Cary Brown, which then became known as the Brown tax (Mooij and Klemm 2021). The brown tax approach looked at project cash flow, the cash inflows and cash outflows. In the case of negative cash inflows (including initial capital investment), the government paid at a certain tax rate to the negative cash flows to the investor.

In 1975, Garnaut and Clunies Ross introduced the Resource Rent Tax (RRT), which applied in the variance of the implementation of the Brown Tax. In this variance, the government no longer contributed to the negative cash flow at a certain tax rate. Instead, the investor was allowed to carry forward the loss into the future with interest. The Australian PRRT is based on Garnaut and Clunies Ross's (1975) RRT regardless of the concerns on neutrality and the efficiency test.

The PRRT was initially designed for oil projects, but it has subsequently been applied for LNG projects more recently in Australia, which currently dominates the industry (Callaghan 2017). PRRT was designed to evade the practical disadvantages of cash flow taxation. Unlike the Brown Tax, the PRRT only taxed positive cash flow and not negative cash flows. It applied to a return after all costs deduction (exploration, development, and production costs). It is very much a way of taxing the profitability of a petroleum project, while the tax losses are carried forward to the future and offset against positive cash flow. Therefore, in 1988, the Australian government introduced the PRRT as they also saw that the PRRT was a more efficient and reasonable fiscal regime for the petroleum industry (Australian Government 2016). However, there was a reduction in PRRT revenue from 2002/2003 to 2015/2016 which then led the Australian government to announce a federal review of the PRRT and related resource taxes in late 2016 (Kraal 2017). The review would provide an assessment of PRRT implementation, considering justice and equitable benefits to the Australian community from the industry without discouraging investments (Callaghan 2017; Kraal 2017). Callaghan (2017) wrote in the Final Review in 2017 that the decreased revenue on PRRT was due to the decline of the production in mature projects, oil and gas price fluctuations, and increases in the deductible expenditures from new projects' large investment and transferable exploration expenditure between companies under the same group. While the PRRT revenue decreased, the LNG production increased. However, this does not indicate that the Australian community is not receiving equitable benefits from the projects (Callaghan 2017).

The Australian government published a report on 'Petroleum Resource Rent Tax Review' in 2017 by Mike Callaghan. A media report released by Frydenberg in late 2018 on the government's responses to the Callaghan PRRT review stated that the review found that "changes should be made to PRRT arrangements to make them more compatible with the developments that have taken place in the Australian oil and gas industry" (Frydenberg 2018). The PRRT was still the preferred way to provide equitable benefits to the Australian community and still attracted the necessary investment. The changes were introduced in July 2019[3] as follows:

- **Lower uplift rates**: The uplift rate on exploration expenditure will be reduced from Long-Term Bond Rate (LTBR)+15 percentage points to LTBR+5. All existing investments will be respected.
- **Onshore projects removed from the PRRT regime (include CBM)**: There has been no PRRT revenue generated since it was first implemented, and this is expected to remain unchanged while it has been used to transfer exploration expenditures to profitable offshore projects and to reduce the PRRT payable. Therefore, this change is expected to strengthen its integrity while simplifying the system.

[3] The changes are not being adopted in this discussion.
Treasury Laws Amendment (2019 Petroleum Resource Rent Tax Reforms No. 1) Act 2019.

- **Review of Gas Transfer Pricing Regulations:**[4] The Treasury commenced a review on this regulation and requested thoughts from industry and communities to contribute to the process of the review.

The three fiscal regimes in the comparison countries, including the technical aspects of CBM, have been discussed. Understanding the technical point of view will provide background on how technical issues impact the fiscal regime (this will be discussed further in Chap. 4).

1.5 Energy Transition and Energy Justice

The energy transition is a pathway towards transformation of the global energy sector from fossil-based to zero-carbon (IRENA 2020). Energy transition refers to the global energy sector's shift from fossil-based systems of energy production and consumption; and this includes oil, natural gas, and coal to RE sources such as wind and solar (S&P Global 2020). The global movement of the energy transition is to support the Paris Agreement that has been signed and ratified by 195 countries. The Paris Agreement is an agreement within the United Nations Framework Convention on Climate Change (UNFCCC) with the main aim to strengthen the global response to the threat of climate change. The Paris Agreement encourages and drives countries to keep the global temperature rises below 2 degrees Celsius and push further efforts to limit the temperature increase to 1.5 degrees Celsius (United Nations 2020a). This movement has raised concerns of justice scholars, in particular on the justice issues in the process of transitioning.

Energy injustices cause and contribute significantly to global inequality as the sector is prone to corruption, taxation issues, environmental damage, GHG emissions, financial issues, and so on (Jenkins et al. 2016). Jenkins et al. (2016) summarised the global energy challenges as being

[4] Twenty-one submissions on the review of Gas Transfer Pricing Regulations were received from expected parties, industry and community (2019): https://treasury.gov.au/consultation/c2019-t364690

created by resource scarcity and population growth in an increasingly unpredictable social and environmental climate. They further explored specific energy justice issues and identified them as related to distributional, recognition, and procedural justice. They define distributional justice as the location in which energy injustices happen in the world. For example, the location of coal mining activities where trucks are the means of transport raised justice concerns among nearby communities, for example in Kalimantan, Indonesia (Sumarno 2020). Recognition-based justice is very much related to sections/groups of society that are ignored/misrepresented in the energy movement (Jenkins et al. 2016). According to Schlosberg (2003), recognition justice aims to fairly represent individuals and free them from any physical threats and offer them complete and equal political rights. Procedural justice is more related to involving communities into the decision-making process, in which policymakers are the main players in the process (Jenkins et al. 2016).

A theory on the just energy transition by Heffron and McCauley (2018) sets out that any efforts moving towards a low-carbon economy must be done in a "just" way for environment and society, and this is what this book focuses on. Further, this book reveals issues regarding CBM development in Indonesia which emphasise the concept of energy trilemma (Heffron et al. 2018). The energy trilemma in Indonesia covers energy and environmental sustainability, energy security, and energy affordability. Heffron and McCauley (2017) discussed further on restorative justice in the energy sector. Restorative justice is primarily used in criminal law. Instead of focusing on punishing the offender, restorative justice aims to repair the harm done to people and/or society/nature. Decision-making would be made taking into consideration the potential harm and the true cost of that decision if restorative justice were applied to the energy sector (Heffron and McCauley 2017). McCauley et al. (2019) stated that their energy justice framework consisted of four elements added with restorative justice: (1) distributional justice; (2) procedural justice; (3) cosmopolitan justice; and (4) recognition justice.

The energy sector is one of the most significant sectors in the world. The importance of managing supply and demand is clear; not only in terms of economics, this sector has significant impacts on society (welfare), environment, health, infrastructure, policy, and many others.

Therefore, energy justice is undoubtedly required in each energy decision and movement, including in the transitioning process from fossil fuels to the full wide-scale development of a just transition to a low-carbon economy.

1.6 Chapters Overview

This book discusses the role of CBM in accelerating the energy transition process in Indonesia. With the aim to look at societal issues, this book also brings an overview on how to reattract investment to CBM projects. The societal issues can bring significant impact to both the local community and the entire nations (Hulme 2002; Bryner 2002; Mudd 2012; Letts 2012). By bringing new perspectives to advance the energy transition in Indonesia, together with other energy sources, will lead to energy sustainability. Further, this book provides the reader with an understanding of the key issues around CBM project development in general and its role in the just transition to a low-carbon economy in Indonesia and beyond.

This book has seven chapters which provide a comprehensive analysis of the Indonesian energy transition movement highlighting how CBM could contribute to Indonesia's energy transition goals.

Chapter 1: Introduction
This chapter provides an overview and context of the book itself, that is, an overview of Indonesia, energy transition, energy justice, CBM, and taxation, by discussing the aims of the book in terms of the energy transition movement particularly for developing and emerging economies.

Chapter 2: Coal Bed Methane and Energy Transition in Indonesia
This chapter covers CBM development in Indonesia a decade on since it started in 2008 and the movement of the country towards a low-carbon economy. This chapter will argue that CBM is known to be a cleaner energy source compared to other fossil fuel such as coal (Bryner 2002; Letts 2012; Flores 2014a), and coal has been the main fuel for power

generation in Indonesia. In 2019, coal power plants accounted for 60% of total electricity generated (National Electricity General Plan Republic of Indonesia 2019). Indonesia has developed its electricity road map towards clean energy and planned to reduce the use of coal by 13% in 2038 (National Electricity General Plan Republic of Indonesia 2019). Therefore, Indonesia should reconsider developing CBM projects with the focus on economic development and bringing fairness and justice to the transition process, helping Indonesia accelerate towards its energy and climate targets. The issues concerning CBM development in Indonesia shall be addressed alongside identifying how investment in CBM projects can drive the energy transition goal as well as improving societal welfare.

Chapter 3: How CBM Projects Align with the Indonesian Energy Transition Goal

This chapter sets out how CBM projects fit into the goal of increasing clean energy supply in Indonesia with regards to the just transition to a low-carbon economy. Having to tackle climate change, Indonesia has started green energy projects through the development of clean energy supplies such as wind power, solar power, bioenergy, geothermal, mini hydro, and hydropower. Considering that CBM is a derivative of fossil fuels, Indonesia has developed CBM under the hydrocarbon regimes. It is important to understand that CBM is also known as one of the cleaner fossil fuel sources that can contribute to the additional clean energy supply for the country. The growth of clean energy projects in Indonesia has increased from year to year and CBM can play an instrumental role in contributing to a cleaner society and environment.

Chapter 4: Current Issues on CBM Development in Indonesia: What Do They Say?

This chapter reveals the real issues regarding the development of CBM projects in Indonesia from different stakeholder perspectives including investors, government, and consultants. Categories and classification of issues are developed and explained in this chapter. It also provides an understanding that there are different ways of seeing a project, yet these issues are interrelated and complex. Furthermore, issues arise due to the impact(s) of other issues, therefore solving issues in isolation will not push CBM development forward. To identify and solve these issues, this chapter will look at two developed countries, the US and Australia, and how they have had successful CBM development and benefited hugely from these projects (Hulme 2002; Bryner 2002; Mackinnon and Fox 2006; Mudd 2012; Letts 2012; Cully 2015).

Chapter 5: Environmental Impacts and Societal Justice Aspect of CBM in Transitioning to Low-Carbon Economy
CBM projects are always located onshore, and this chapter discusses how CBM projects have impacted local communities both economically and environmentally. A survey on the local impacts and local community's expectation was conducted to collect the data for this purpose. This chapter will also elaborate further in terms of energy justice issues which will be explored further in regard to CBM development, for example what do they want from the project apart from energy supplies for the community (i.e. distributive, procedural, recognition, and restorative justice)? Therefore, this chapter explores in detail the impacts of project development on local community development and how the impacts and benefits can be distributed fairly to the community.

Chapter 6: Reattracting CBM Investment in Indonesia
This chapter provides ideas on how countries (with a focus on Indonesia as the case study) can achieve energy transitions goals by reattracting CBM investment. How would the government be expected to be involved in the success of CBM development and support investors to provide equitable benefits to society? Issues are defined earlier (in previous chapters; see above), and this chapter highlights what can be done to solve the issues concerning society and environmental justice through taxation by utilising examples from the US and Australia. Taxation is one of the key tools in the development of any country and is very much related to welfare distribution; therefore, this chapter highlights taxation modifications to improve CBM development and create project attractiveness for governments and investors. Lessons from other countries are drawn from this chapter in the process of suggesting modifications to the taxation system in Indonesia.

Chapter 7: The Future of CBM in Indonesia and International Reflections

This chapter concludes the book, exploring in particular the future outlook of the energy transition in Indonesia and what role CBM can play in energy transition and energy sustainability. As such, it will examine the dos and dont's for the government in developing the clean energy project as a whole and in supporting CBM development in the country. Finally, this chapter provides a number of recommendations to the GoI to achieve future energy transition goals out towards 2050.

References

National Exploration Committee. 2015. Jakarta.

Anderson ZurMuechlen and Co., 2001. *Coalbed Methane Development in the Powder River Basin of Montana: Economic and Social Impacts of Proposed Development.* [online] Montana: The State of Montana Government. Available at: https://deq.mt.gov/Portals/112/Energy/CoalbedMethane/Documents/cbm2.pdf [Accessed 19 March 2020].

Australian Government, 2016. *Petroleum Resource Rent Tax Review.* Commonwealth of Australia.

Bachtiar, A., 2018. *A Focus Group Discussion on CBM Industry in Indonesia.*

Bank, G., and Kuuskraa, V., 2006. *The Economics of Powder River Basin Coalbed Methane Development.* Washington D.C.: Advanced Resources International, Inc.

Bayar, Y., 2017. Greenfield and Brownfield Investments and Economic Growth: Evidence from Central and Eastern European Union Countries. *Naše gospodarstvo/Our economy*, 63(3), pp. 19–26.

Bowe, M. and Moore, T., 2015. Coalbed Methane Potential and Current Realisation in Indonesia. *AAPG*, [online] 90234. Available at: http://www.searchanddiscovery.com/abstracts/pdf/2015/90234gtw/abstracts/ndx_bowe.pdf [Accessed 20 February 2017].

Bryner, G., 2002. *Coalbed Methane Development in the Intermountain West: Primer*. Colorado: University of Colorado Law School.

Calder, J., 2014. *Administering Fiscal Regimes for Extractive Industries.* [ebook] International Monetary Fund. Available at: https://www.elibrary.imf.org/

view/IMF071/20884-9781475575170/20884-9781475575170/20884-9781475575170.xml?language=en&redirect=true [Accessed 21 March 2020].
Callaghan, M., 2017. *Petroleum Resource Rent Tax Review*. Canberra: Australian Government.
Cbmasia.ca. 2016. *CBM ASIA*. [online] Available at: http://www.cbmasia.ca/CBM-In-Indonesia# [Accessed 12 April 2016].
Chakhmakhchev, A. and Fryklund, B., 2008. Critical success factors of CBM development – Implications of two strategies to global development. In: *19th World Petroleum Congress*. USA: IHS.
Collier, P., 2010. Principles of resource taxation for low-income countries. In: P. Daniel, M. Keen and C. McPherson, ed., *The Taxation of Petroleum and Minerals: Principles, Problem, and Practice*. Oxon: Routledge, pp. 75–86.
Cully, M., 2015. *Review of the Socioeconomic Impacts of Coal Seam Gas in Queensland*. Canberra: Australian Government, Department of Industry, Innovation, and Science.
Dewan Energi National Indonesia (National Energy Council of Indonesia), 2020. *National Energy Mix 2020*. Jakarta: National Energy Council – Dewan Energi Nasional Indonesia.
Flores, R., 2014a. *Coal and Coal Bed Gas: Fueling the Future*. United States of America: Elsevier Inc., pp. 567–574.
Flores, R., 2014b. *Coal and Coalbed Gas*. Amsterdam: Elsevier, pp. 2–39.
Garnaut, R. and Ross, A., 1975. Uncertainty, Risk Aversion and the Taxing of Natural Resource Projects. *The Economic Journal*, 85(338), p. 272.
Heffron, R. and McCauley, D., 2017. The concept of energy justice across the disciplines. *Energy Policy*, 105, pp. 658–667.
Heffron, R. and McCauley, D., 2018. What is the 'Just Transition'?. *Geoforum*, 88, pp. 74–77.
Heffron, R., McCauley, D. and de Rubens, G., 2018. Balancing the energy trilemma through the Energy Justice Metric. *Applied Energy*, 229, pp. 1191–1201.
Hulme, D., 2002. *Coalbed Methane Development in Wyoming's Powder River Basin*. [ebook] Colorado: University of Colorado Law School. Available at: http://scholar.law.colorado.edu/coalbed-methane-development-intermountain-west/4 [Accessed 19 March 2020].
International Monetary Fund, 2012. *Fiscal Regimes for Extractive Industries: Design and Implementation*. [online] International Monetary Fund. Available at: https://www.google.com/url?sa=t&rct=j&q=&esrc=s&source=web&cd=3&cad=rja&uact=8&ved=2ahUKEwi6nozFq9HoAhWxQkEAHVs5BTQQ

FjACegQIARAB&url=https%3A%2F%2Fwww.imf.org%2Fexternal%2Fnp%2Fpp%2Feng%2F2012%2F081512.pdf&usg=AOvVaw3%2D%2DNziWcwfWOVklkY88BOl [Accessed 20 March 2020].

International Monetary Fund, 2015. *"Options for Low Income Countries' Effective and Efficient Use of Tax Incentives for Investment."* [online] International Monetary Fund. Available at: https://www.imf.org/external/np/g20/pdf/101515.pdf [Accessed 20 March 2020].

IRENA, 2020. *Energy Transition.* [online] Irena.org. Available at: https://www.irena.org/energytransition [Accessed 20 July 2020].

Jenkins, K., McCauley, D., Heffron, R., Stephan, H. and Rehner, R., 2016. Energy justice: A conceptual review. *Energy Research & Social Science*, 11, pp. 174–182.

Johnson, C., 1990. Ranking countries for minerals exploration. *Natural Resources Forum*, 14(3), pp. 178–186.

Kasatuka, C. and Minnitt, R., 2006. Investment and non-commercial risks in developing countries. *The Journal of the Southern African Institute of Mining and Metallurgy*, 106(12), pp. 849–856.

Kraal, D., 2017. Review of Australia's Petroleum Resource Rent Tax: Implications from a Case Study of the Gorgon Gas Project. *Federal Law Review*, 45(2), pp. 315–349.

Kunce, M., Gerking, S., Morgan, W. and Maddux, R., 2003. State Taxation, Exploration, and Production in the U.S. Oil Industry. *Journal of Regional Science*, 43(4), pp. 749–770.

Kusumah, G. and Munawar, A., 2015. How to Make Coal Bed Methane Work in Indonesia. In: *Indonesian Petroleum Association Conference*. Jakarta: Association for American Petroleum Geologist.

Lalean, B., 2008. Designing a Production Sharing Contract (PSC) for Coal Bed Methane Project in order to Balance the Shares Portions of the Government and the Contractor. In: *Indonesian Petroleum Association Conference*. Jakarta: Association for American Petroleum Geologist.

Lalean, B., 2010. Launching a First Coal Bed Methane Project in Indonesia – A Case Study. In: *Indonesian Petroleum Association Conference*. Jakarta: Association for American Petroleum Geologist.

Lalean, B., 2012. Seeking Workable CBM Business Model in Indonesia. In: *Indonesian Petroleum Association Conference*. Jakarta: Association for American Petroleum Geologist.

Len, C., 2014. *Indonesia Coalbed Methane*. Bountiful Promise Help Up by Institutional Paralysis. [online] Singapore: Energy Studies Institute. Available at: http://esi.nus.edu.sg/docs/default-source/esi-policy-briefs/indonesia-cbm-bountiful-promise-held-up-by-institutional-paralysis.pdf [Accessed 28 March 2016].

Letts, L., 2012. *Coal Seam Gas Production – Friend Or Foe of Queensland's Water Resources?*. 1st ed. Australia: Thomson Reuters.

MacKinnon, A. and Fox, K., 2006. Demanding Beneficial Use: Opportunities and Obligations for Wyoming Regulators in Coalbed Methane. *Wyoming Law Review*, [online] 6(2), pp. 369–400. Available at: https://scholarship.law.uwyo.edu/cgi/viewcontent.cgi?article=1135&context=wlr [Accessed 19 March 2020].

Makmun and Sitepu, E., 2012. Makmun. *Effect of fiscal incentives on coal bed methane price: a hypothetical analysis*, 3(1).

McCauley, D., Ramasar, V., Heffron, R., Sovacool, B., Mebratu, D. and Mundaca, L., 2019. Energy justice in the transition to low carbon energy systems: Exploring key themes in interdisciplinary research. *Applied Energy*, 233–234, pp. 916–921.

Metters, C., 2014. Delivering Value in PSC Environment. In: *Indonesian Petroleum Association, 38th Annual Convention & Exhibition*. Jakarta: Indonesian Petroleum Association.

Mooij, R. and Klemm, A., 2021. Why and How to Tax Corporate Income. In: R. Mooij, A. Klemm and V. Perry, ed., Corporate Income Taxes under Pressure. [online] International Monetary Fund, p. 388. Available at: https://www.elibrary.imf.org/view/books/071/28329-9781513511771-en/ch002.xml [Accessed 13 July 2021].

Morgan, P., 2002. Mineral title management—the key to attracting foreign mining investment in developing countries?. *Applied Earth Science*, 111(3), pp. 165–170.

Morisset, J., 1999. *Foreign Direct Investment in Africa: Policies Also Matter*. World Bank Policy Research Working Paper, No. 2481. [online] World Bank. Available at: http://papers.ssrn.com/sol3/papers.cfm?abstract_id=632551 [Accessed 20 March 2020].

Mudd, M., 2012. Montana v. Wyoming: An Opportunity to Right the Course for Coalbed Methane Development and Prior Appropriation. *Golden Gate University Environmental Law Journal*, 5(2), pp. 297–342.

National Electricity General Plan Republic of Indonesia, 2019. *National Electricity General Plan Republic of Indonesia*. Jakarta: Ministry of Energy and Mineral Resources.

O'Neill, D., 1993. Mining investment in Africa. *Natural Resources Forum*, 17(4), pp. 262–272.

Otto, J., 1992. *A Global Survey of Mineral Company Investment Preferences. In Mineral Investment Conditions in Selected Countries of the Asia-Pacific Region*. New York: United Nations.

Penney, K., McCallum, R., Schultz, A. and Ball, A., 2007. *Mineral Exploration In APEC Economies: A Framework for Investment*. Australian Bureau of Agricultural and Resource Economics (ABARE) research report, December. [online] Australian Bureau of Agricultural and Resource Economics (ABARE). Available at: http://publications.apec.org/publication-detail.php?pub_id=244 [Accessed 20 March 2020].

PricewaterhouseCoopers, 2014. *Oil and Gas in Indonesia*. Investment and Taxation Guide, May 2014 – 6th edition. Jakarta: PricewaterhouseCoopers.

PricewaterhouseCoopers, 2016. *Oil and Gas in Indonesia*. Investment and Taxation Guide, May 2016 – 7th edition. Jakarta: PricewaterhouseCoopers.

Redhead, A., 2018. *Tax Incentives in Mining: Minimising Risks to Revenue*. [ebook] Canada: International Institute for Sustainable Development. Available at: https://www.google.com/url?sa=t&rct=j&q=&esrc=s&source=web&cd=2&cad=rja&uact=8&ved=2ahUKEwjvmofc-cnoAhWpQxUIHQeZDmYQFjABegQIBBAB&url=https%3A%2F%2Fwww.oecd.org%2Ftax%2Fbeps%2Ftax-incentives-in-mining-minimising-risks-to-revenue-oecd-igf.pdf&usg=AOvVaw02chYk5AZ00FBSPJnbvrf3 [Accessed 20 March 2020].

S&P Global, 2020. *What Is Energy Transition?*. [online] Spglobal.com. Available at: https://www.spglobal.com/en/research-insights/articles/what-is-energy-transition [Accessed 20 July 2020].

Schlosberg, D., 2003. The justice of environmental justice: Reconciling equity, recognition, and participation in a political movement. In: A. Light and A. De-Shalit, ed., *Moral and political reasoning in environmental practice*. Cambridge, Mass.: MIT Press.

Sigam, C. and Garcia, L., 2012. *Extractive Industries: Optimizing Value Retention in Host Countries*. [ebook] Geneva: United Nations Conference on Trade and Development. Available at: https://unctad.org/en/PublicationsLibrary/suc2012d1_en.pdf [Accessed 23 March 2020].

SKKMigas, 2018. *Coal Bed Methane Blocks in Indonesia 2008–2018*. Jakarta: SKKMigas.

Skov, J. and Myers, N., 2004. *Easy Money, Hidden Costs: Applying Precautionary Economic Analysis to Coalbed Methane in the Powder River Basin.* [online] Iowa: Science and Environmental Health Network. Available at: http://www.sehn.org/tccpdf/coalbed%20methane%20costs%20Powder%20River.pdf [Accessed 16 September 2017].

Stevens, H., Scott, H., Sani, K. and Hardjosuwiryo, S., 2001. Indonesia's 337 TCF CBM Resource a Low Cost Alternative to Gas, LNG. *Oil and Gas Journal*, (October 22).

Stevens, S. and Hadiyanto, 2004. Indonesia: Coalbed Methane Indicators and Basin Evaluation. In: *SPE Asia Pacific Oil and Gas Conference and Exhibition.* Society of Petroleum Engineers.

Sukhyar, R. and Fakhruddin, R., 2013. *Unconventional Oil & Gas Potential in Indonesia With Special Attention to Shale Gas and Coal Bed Methane.*

Sumarno, T., 2020. *Identifying and Modifying the Coal Bed Methane Fiscal Regime in Indonesia: A Comparative Study with the Australia and the U.S.. Doctorate*. University of Dundee.

The Hon Josh Frydenberg MP, Treasurer of the Commonwealth of Australia, 2018. *Government Responds to Callaghan PRRT Review*. [online] Available at: https://ministers.treasury.gov.au/ministers/josh-frydenberg-2018/media-releases/government-responds-callaghan-prrt-review [Accessed 26 March 2020].

Tienhaara, K., 2006. Mineral investment and the regulation of the environment in developing countries: Lessons from Ghana. *International Environmental Agreements: Politics, Law and Economics*, 6(4), pp. 371–394.

Tole, L. and Koop, G., 2010. Do environmental regulations affect the location decisions of multinational gold mining firms?. *Journal of Economic Geography*, 11(1), pp. 151–177.

Treasury Laws Amendment (2019 Petroleum Resource Rent Tax Reforms No. 1) Act 2019. Royal Assented to 5 April 2019.

United Nations, 2017. *United Nations Handbook on Selected Issues for Taxation of the Extractive Industries by Developing Countries.* [online] New York: United Nations. Available at: https://www.google.com/url?sa=t&rct=j&q=&esrc=s&source=web&cd=1&cad=rja&uact=8&ved=2ahUKEwj9x7LDitHoAhVEr3EKHTqWCuMQFjAAegQIBRAB&url=https%3A%2F%2Fwww.un.org%2Fesa%2Fffd%2Fwp-content%2Fuploads%2F2018%2F05%2FExtractives-Handbook_2017.pdf&usg=AOvVaw3yELQwyKiQlkB11UPh1Vo9 [Accessed 23 March 2020].

United Nations, 2020a. [online] Unfccc.int. Available at: https://unfccc.int/process-and-meetings/the-paris-agreement/the-paris-agreement [Accessed 21 July 2020].

United Nations, 2020b. *All Ndcs*. [online] Www4.unfccc.int. Available at: https://www4.unfccc.int/sites/NDCStaging/Pages/All.aspx [Accessed 9 November 2020].

Vivoda, V., 2017. Determinants of Foreign Direct Investment in the Mining Industry. *Mining in the Asia-Pacific*, pp. 19–33.

Waeckerlin, E., 2005. The New Border War: CBM Development in the Powder River Basin of Wyoming and Montana. *Public Land & Resources Law Review*, 26, pp. 149–162.

WoodMackenzie, 2015. *Sumatra CBM*. United Kingdom: Wood Mackenzie.

World Bank, 2020a. *Indonesia | Data*. [online] Data.worldbank.org. Available at: https://data.worldbank.org/country/ID [Accessed 11 November 2020].

World Bank, 2020b. *Overview*. [online] World Bank. Available at: https://www.worldbank.org/en/country/indonesia/overview [Accessed 11 November 2020].

Worldometers.info. 2020. *Indonesia Population (2020) – Worldometer*. [online] Available at: https://www.worldometers.info/world-population/indonesia-population/ [Accessed 6 July 2020].

2

Energy Transition and Coal Bed Methane Development in Indonesia

Abstract In the next chapter, the book discusses CBM development in Indonesia a decade on since it started in 2008 and the movement of the country towards a low-carbon economy. This chapter will argue that CBM is known to be a cleaner energy source compared to other fossil fuel such as coal (Bryner, *Coalbed Methane Development in the Intermountain West: Primer*. Colorado: University of Colorado Law School, 2002; Letts, *Coal Seam Gas Production – Friend Or Foe Of Queensland'S Water Resources?*. 1st ed. Australia: Thomson Reuters, 2012; Flores, *Coal and Coal Bed Gas: Fueling the Future*. United States of America: Elsevier Inc., pp. 567–574; *Coal and Coalbed Gas*. Amsterdam: Elsevier, pp. 2–39, 2014), and coal has been the main fuel for power generation in Indonesia. In 2019, coal power plants accounted for 60% of total electricity generated (National Electricity General Plan Republic of Indonesia, *National Electricity General Plan Republic of Indonesia*. Jakarta: Ministry of Energy and Mineral Resources, 2019). Indonesia has developed its electricity road map towards clean energy and planned to reduce the use of coal by 13% in 2038 (National Electricity General Plan Republic of Indonesia, *National Electricity General Plan Republic of Indonesia*. Jakarta: Ministry of Energy and Mineral Resources, 2019). Therefore, Indonesia should

T. B. Sumarno, *Just Energy Transitions and Coal Bed Methane*, Energy, Climate and the Environment, https://doi.org/10.1007/978-3-030-85490-4_2

reconsider developing CBM projects with the focus on economic development and bringing fairness and justice to the transition process, helping Indonesia accelerate towards its energy and climate targets. The issues concerning CBM development in Indonesia shall be addressed alongside identifying how investment in CBM projects can drive the energy transition goal as well as improving societal welfare.

Keywords Australia • Coal bed methane • Energy development • Energy transition • Environment • Exploration • Fiscal regime • Fiscal incentives • Indonesian • Justice • Petroleum Resource Rent Tax (PRRT) • Energy policy • Powder River Basin • Production sharing contract (PSC) • Regulations • Resource rent • Royalty • Society • Taxation • The US • Wyoming

2.1 Introduction

This chapter covers CBM development in Indonesia since it started in 2008 and the movement of the country towards a low-carbon economy. In conjunction with the transition, it is important to understand that Indonesia is one of the most populated countries in the world with 274 million people by 2020, and projected to grow and reach 305 million by 2035. A variety of energy projects are needed across the country due to rising energy demand. In regard to clean energy project developments, there is a lack of understanding for the local environment among national policy makers in Indonesia. There is also a need to have a more comprehensive and meaningful consultation between decision-makers and industry players as well as other stakeholders (e.g. society) in the process of formulating energy policy.

CBM projects are known to be a cleaner energy source of energy in comparison to other fossil fuels such as coal (Bryner 2002; Letts 2012; Flores 2014a). Coal has been the main fuel for power generation in Indonesia. In 2019, 60% of total electricity was generated by coal power plants (National Electricity General Plan Republic of Indonesia 2019). However, Indonesia has developed its electricity road map towards clean energy and planned to reduce the use of coal by 13% in 2038 (National

Electricity General Plan Republic of Indonesia 2019). CBM resources are onshore and close or in local communities' areas where having energy project in the areas could benefit and improve energy access in these communities. Therefore, Indonesia should reconsider developing CBM projects with a focus on economic development and bringing fairness and justice to the energy transition. The CBM is actually natural gas, which could be used to replace coal in electricity power plant or could be used directly for locally developed gas power plant for the community. Natural gas is known to have lower emission than coal. Which is why this will help Indonesia accelerate towards its energy and climate target. The issues concerning CBM development in Indonesia shall be addressed, identifying how investment in CBM projects can drive the energy transition goal as well as improving societal welfare and increasing domestic energy supply.

2.2 Energy Transition in Indonesia

Indonesia published its first blueprint on climate change mitigation and adaptation in 2008 (Indonesian Fiscal Policy Office 2008), and recently committed to reducing the GHG emission by 29% by 2030 under a "business as usual" case (Ministry of Finance, Republic of Indonesia 2020; United Nations 2020a, b).

As one of the Association of South East Asian Nations (ASEAN) member states, Indonesia, along with the nine other countries, has committed and agreed to increase the share of new and renewable energy sources into their energy mixes to 23% by 2025. For Indonesia, the 23% comprises CBM, shale gas, and RE sources[1] such as biomass, biofuel, solar PV, and wind power (ASEAN Centre for Energy 2015). Although shale gas and CBM could be categorised as 'new', they are definitely not renewable, neither do they have the attributes of many RE sources (clean, etc.). The Asian Development Bank (2020) published a report on incentives for RE projects in Indonesia, such as fiscal incentives (tax holiday) and subsidies (direct and indirect subsidies). According to the Ministry of Finance

[1] Indonesia considers biomass and biofuel as RE sources.

(2020), the GoI formulated strategy to achieve this goal by accelerating the following actions:

1. Establishing a Public Service Agency that specialises in managing environmental funds in the context of climate change with the Minister of the Environment (KLH);
2. In the National Income and Budget, funds have been allocated for matters relating to climate change or what is called Climate Budget Tagging (CBT);
3. Producing the world's first Green Sukuk,[2] which will later be tied to Indonesian projects and programmes to reduce pollution; and
4. Providing incentives for projects that are committed to being green, renewable, geothermal energy and others taken from the National Income and Expenditures Budget (APBN).

Aligned with the Ministry of Finance strategy, the MoEMR also formulated its strategy to achieve this goal by reducing the supply and demand of fossil fuel especially for oil and coal and increasing the supply and demand of RE as well as natural gas. Indonesia is known to have a total potential of 442 GigaWatts (GW) of RE with current installed capacity amounting to 9.32 GW or 2% of the total potential (Directorate General New and Renewable Energy 2016). This potential RE consists of solar PV (207.9 GW), hydropower (75 GW), wind power (60.6 GW), bioenergy (32.6 GW), geothermal (38.5 GW), ocean (tidal) energy (17.9 GW), and mini hydropower and micro hydropower (19.4 GW).

Both the government scholars (Kennedy 2018; Ialnazov and Keeley 2020) and some other international institutions have paid significant attention to the growth of the RE projects in the country, and the strategy to slowly move away from the fossil fuels. However, regardless of its target to increase the new energy (of which CBM is a part of) and RE in the energy mix, there is distinct lack of concern regarding CBM development in Indonesia to contribute to the 23% target by 2025.

---

[2]Green Sukuk are Shari'ah compliant investments in renewable energy and other environmental assets. They address Shari'ah concern for protecting the environment.

Moreover, despite having so much potential in both fossil and RE, Indonesia has failed to provide full access to electricity for the whole country by 2020 (IESR 2019). Progress on RE development in Indonesia has also not been as fast as expected. While speeding up the RE development in Indonesia, other less carbon emission intensive energy resources such as gas would also enhance the energy transition in Indonesia. Indonesia has 453 TCF of CBM, which has been suggested to be used for electrifying regions where they are located. Developing an electricity grid for CBM could also be used for renewable energy, thus helping to lower the cost of RE and create a more economic project for all of them as they would use integrated infrastructure (facilities sharing).

Indonesia aims to become climate neutral by 2060 (Ministry of Energy and Mineral Resources 2021). One of the strategies in the energy sector is by slowly phasing out its coal-fired power plant and replacing them with the RE sources. With another target 35GW by 2030, Indonesia aims to increase its coal-fired power plant until 2035, and only start to phase them out thereafter. A better option of achieving 35GW target is by increasing gas power plant, rather than coal power plant, as it has less emission. CBM projects have been started since 2008, and many of them were terminated, Indonesia could reattract the investors to continue their projects by giving fiscal incentives. CBM could be an option to accelerate the energy transition, instead of building up coal-fired power plants for the next 15 years.

2.3 CBM as a Cleaner Source of Energy

CBM is known to be a cleaner energy source compared to other fossil fuels such as coal (Bryner 2002; Letts 2012; Flores 2014a). Although coal has historically been the main fuel for power generation, Indonesia, through its electricity road map, has already shifted towards using clean energy and has also planned to reduce coal use by 13% in 2038 (National Electricity General Plan Republic of Indonesia 2019).

The use of CBM also demonstrates its value through a two-pronged reduction in GHGs. First, with CBM, the highly active GHG methane is transformed into less harmful CO_2 even prior to its release during coal mining (methane released into the atmosphere has a 21-fold greater climate relevance than CO_2). Second, CO_2 emissions per kWh can be reduced markedly by substituting fossil fuels. In addition to the ecological benefits, this also results in an economic advantage, since related projects qualify for carbon credits (Clarke Energy 2020). In India, CBM is one of the emerging clean coal technologies for the recovery of energy by unconventional methods (Mallick and Prabu 2017).

CBM is also known as a 'new' energy in the Indonesian energy mix; however, it is not being targeted specifically in the NEP (Government Regulation 79/2014). According to the Review into the NEP achievement, currently Indonesia has only reached 9.5% of its 23% target of new and RE in the 2025 energy mix target (National Energy Council 2020). CBM, as a cleaner energy source than other fossil fuels, could align with the Indonesian NEP target 2025.

CBM is gas (methane) trapped in coal structures, and extracting the methane from the coal will require specific technology as explained earlier. CBM as a natural gas has low air pollutant emissions, much lower than that of coal (approximately 50%). The use of CBM for electrification could potentially lead to rapidly improving air quality when substituting other combustible fuels such as coal/oil. As Indonesia has attempted to reduce the use of coal as power plant fuel, replacing coal with CBM (gas) will bring CO_2 emissions down by around 40% for each unit of energy output (International Energy Agency 2019).

2.4 Understanding CBM: Technical and Geological Perspective

The primary differences between the CBM reservoirs and the conventional gas are water production and the gas storage mechanism (Al-Jubori et al. 2009). Al-Jubori et al. (2009) identified fundamental challenges in CBM production, including economic, geological, logistical and operational challenges, and dewatering management has become the primary consideration. Flores (2014b) stated that CBM is distinct from conventional gas due to its generation, gas storage capacity, its migration, and its entrapment within the coal.

Most conventional gas storage capacity is related to porosity as it is trapped and stored in the pore systems of the matrix. Coal can store more gas than an equivalent volume of sandstone at a similar pressure even though they have moderate intrinsic porosity. A coal's rank determines the gas storage capacity; higher-rank coals have a better gas storage capacity. This is only one of the requirements for a project to be commercially successful. The methane in high-rank coals is generated during the thermal maturation of the organic compounds, while the methane in low-rank coals is generated by microbial activity (Al-Jubori et al. 2009).

Coalification is the process of transforming organic-rich peat into coal through biological processes, and the application of heat and pressure (Al-Jubori et al. 2009). Gas storage capacity is determined by its coal rank and pressures. A higher rank of coal and higher pressures will thus have a better gas storage capacity. The highest rank of coal is called anthracite, which has the highest gas storage capacity, followed by various grades of bituminous, which are bituminous, sub-bituminous, and lignite as the lowest rank. The methane is generated by microbial activity and by heat. The organic materials that form the coal adsorb the methane on the surface, and the biogenetic processes stop since these materials are transformed to higher-rank coal and are exposed to more heat.

Allowing gas to flow from the coal seams to the wellbore requires permeability, in which the natural fracture system in coal is the principal mechanism for permeability. This may be enhanced by tectonic stress, although excessive tectonic activity leads to a reduction in permeability (Al-Jubori et al. 2009). This also involves the dewatering process to create a lower pressure in the reservoir, thus increasing the gas permeability of the coal, and thus facilitating the gas to migrate to the wellbore.

As previously discussed, dewatering management is important in the CBM industry. The water produced during the CBM production must be disposed of either by reinjecting the water into a deeper zone or by discharging it on the surface following water treatment. The water being discharged at the surface must be treated to ensure that it is safe for the environment and community, and the gas is separated from the water between the tubing and casing.

CBM wells are generally characterised by low production rates. Many CBM wells are stimulated by hydraulic fracturing to connect the cleats and natural fractures with the wellbore to contact the maximum drainage area. Some CBM basins have high natural permeability, greater than 100 mD, and do not require stimulation (Al-Jubori et al. 2009).

There are exceptions to the dewatering model. Some wells produce gas immediately, without the lengthy dewatering process. Mature fields may be partially or even fully dewatered as a result of previous production. This is similar to the case for wells completed in coal seams where water has been removed during the mining operations. Horizontal boreholes are often drilled into coal seams prior to mining to reduce the methane level. In a similar approach, drilling horizontal wells creates highly effective conduits for CBM production (Li and Zhang 2016). Some areas, such as the San Juan Basin in the US, can produce gas without stimulation at volumes and rates comparable to production from conventional sandstone reservoirs (Al-Jubori et al. 2009), whereas in other coal-producing basins, multilateral wells are placed within coal seams to maximise production.

2.5 CBM Worldwide and CBM Outlook

Australia started the production of CBM in 2008 with 4 billion m^3 (141 bcf), and China produced more than 1.4 million m^3 in 2006. These volumes were very small compared to US production in 2007, which was 61 billion m^3 (2.15 tcf). The main driving factor that triggered the development of CBM in the US in the 1980s was that of tax incentives (Al-Jubori et al. 2009).

Al-Jubori et al. (2009) noted in a review on 'Coal Bed Methane as a Clean Energy of the World' some countries that potentially produce the CBM in these past few years, namely the US, Australia, China and India. Although Indonesia was not mentioned as one of the countries potentially producing CBM, Indonesia also evidenced CBM potential, known as it is as one of the countries with significant coalbed gas projects (Flores 2014b). Australia has also been predicted to be a significant contributor to LNG exports of gas from its CBM (Cassidy and Kosev 2015).

The world has been very concerned about the environment. This leads to the need for clean energy supplies and to reduce GHG emissions (Flores 2014b). "Coal bed methane (CBM) is an unconventional form of natural gas found in coal deposits or coal seams. It is a primary clean energy source of natural gas" (Haldar 2018). According to IPCC Special Report on Carbon dioxide Capture and Storage (n.d.), the emission factor of natural gas is half of coal's emission factor (IPCC, n.d.). While coal has 87–97 gCO_2/MJ, natural gas only has 50 gCO_2/MJ, and in fact gas has the lowest emission factor amongst all type of fossil fuels (IPCC n.d.). It may not be cleaner than the RE in general, but given an abundant resources on CBM, Indonesia could develop more of less emission fossil fuel rather than keep developing higher polluting energy sources. Indonesia of course still has to accelerate its RE development while redirecting its support from coal/oil to CBM and RE. Indeed, CBM has been known as a clean energy source (Flores 2014a; Letts 2012); in other words, CBM will gradually grow following the concern about world energy needs: "The U.S. and Canada produce sizeable coalbed gas and Australia is the most advanced major producer of coalbed gas outside the US" (Flores 2014b).

CBM could become a stable, economically viable source of natural gas, especially for the Non-Organisation for Economic Cooperation and Development (NOECD) Asian countries, which includes Indonesia (Flores 2014b). This project tends to be an inexpensive source of clean energy (Flores 2014b). One of the reasons that coalbed gas has not been significantly extracted is because the countries with reserves have not been giving enough incentives to drive exploration and development activities (Flores 2014a). It is clear that the incentives, which are part of the fiscal regimes, has a significant influence on exploration and development activities (Calder 2014; Redhead 2018; IMF 2015; Inter-Governmental Forum – OECD 2020). In consequence, it has been a concern of the GoI to apply the proper fiscal regimes to CBM projects.

2.6 CBM in Indonesia

2.6.1 The CBM History in Indonesia

In 1998, Indonesia enacted Ministerial Decree on the Implementation of Coal Bed Methane Development, that is, the Ministerial Decree No. 1669.K/30/MEP/1998 (see Fig. 2.1). Under this decree, the Research and Development Department of Lemigas, the Oil and Gas Research Institution of the Republic of Indonesia, initially conducted research on CBM in 2002. According to Mrs. Legowo, the then chairwoman of the institution, CBM in Indonesia was considered as one of the energy resources never explored before. In 2006, the minister of energy and mineral resources commanded the laboratory to stop research and start fieldwork for CBM. Medco, a national oil and gas company, signed a memorandum of understanding (MOU) with the institution in 2005, agreeing to do the project together (Legowo 2018; Rosyadi 2010). The budget was from the government, and the field was Medco's. The fieldwork was completed at the end of 2005, and by 2006 there was already a CBM rig in Rambutan field, located in the South Sumatra Basin. At that time, the result was considered good enough to kick off the CBM project. The draft regulation had been prepared since 2004, and in 2006 Mrs

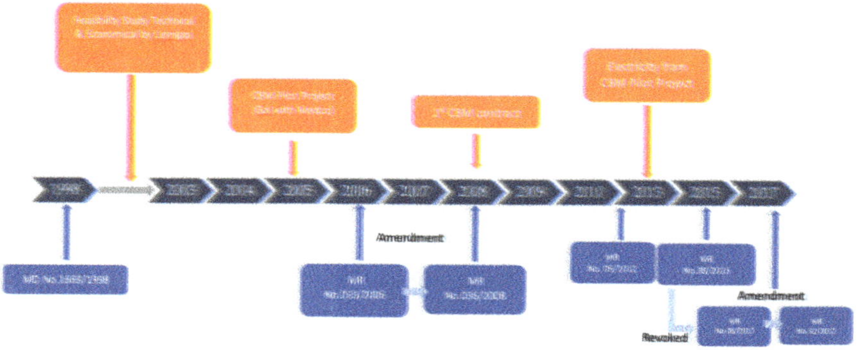

Fig. 2.1 Road map of CBM development in Indonesia. (Source: Created and adapted by the author (2020 from Siirait 2013))

Legowo was appointed as the special staff of the minister until 2008. Subsequently, the minister of energy and mineral resources enacted Ministerial Regulation No. 33/2006 to regulate the CBM development activities which then was amended in 2008 by MR No. 36/2008. This MR became the basis for awarding the first CBM contract in 2008 (Legowo 2018).

The following part discusses the current situation of CBM in Indonesia from different perspectives as a result of a series of interviews and focus group discussions conducted in early 2018.[3]

2.6.2 CBM Resources in Indonesia

Indonesia's CBM resource is located in the archipelago basins with total potential CBM resources of 453 Tcf (see Table 2.1) (General Plan of

Table 2.1 Indonesia's CBM potential resources

No	Basin	Province	Average depth (m)	CBM resources (TCF)
1	South Sumatra	South Sumatra	762	183
2	Barito	South Kalimantan	915	102
3	Kutai	East Kalimanan	915	80
4	Central Sumatra	Riau	762	53
5	North Tarakan	East Kalimantan	701	18
6	Berau	East Kalimantan	671	8.4
7	Ombilin	West Sumatra	762	0.5
8	Pasir Asem	South Kalimantan	701	3.0
9	North West Java	West Java	1524	0.8
10	Sulawesi	South Sulawesi	610	2.0
11	Bengkulu	Bengkulu	610	3.6
Total CBM resources				453

Source: General Plan of National Electrification (2019); Stevens and Hadiyanto (2004)

[3] Author's research in 2017–2020.

National Electrification 2019; Wood Mackenzie 2015; CBM Asia 2016; Stevens et al. 2001; Stevens and Hadiyanto 2004). About 85% of these basins contain lignite to sub-bituminous (low-rank coal), with the remaining 15% bituminous to anthracite (low-rank coal) (Flores 2014a: 568). CBM operators thought that Indonesian CBM was not a prospective resource due to its rank until the success of the PRB in the US (Stevens and Hadiyanto 2004). The sub-bituminous coal has similarities with the PRB, the largest coal bed basin in the US (Moore et al. 2014; Davis et al. 2007; Stevens and Hadiyanto 2004) in terms of coal thickness, coal rank, area extent, and gas adsorption capacity (Flores 2014a). The Miocene coal deposits in Indonesia have a thicker, deeper, and higher rank of coal than the PRB (Stevens and Hadiyanto 2004), and therefore Indonesia is expected to have better resources.

Len (2014) noted an optimism about CBM long-term opportunities and development in Indonesia, given that it was recognised as a prospective resource. First, as the second largest CBM resource in Asia after China (CBM Asia official website, www.cbmasia.ca 2016), Indonesia has a favourable geological condition for CBM production. Second, there is an increase in natural gas domestic needs. CBM is mainly used to fuel power plants. As the need for electricity in Indonesia (see Fig. 2.2) is forecasted to increase 4.67 times in 2019 from 2015 (Tharakan 2015), CBM option is better than coal. Third, CBM basins are mostly located in Sumatra and Kalimantan, which already possess gas operations and infrastructure in place that support the delivery process for domestic needs (International Energy Agency 2014; Directorate General of Electricity Ministry of Energy & Mineral Resources, Republic of Indonesia 2017). Fourth, the GoI seemed to support the CBM exploration in Indonesia. This has been strengthened by the committee constituted by the MoEMR, the National Exploration Committee (NEC). Fifth, it was known that the exploration costs for CBM in Indonesia were lower than those for conventional onshore drilling.

Indonesia is expected to start developing 6 out of 11 basins: South Sumatra Basin, Barito Basin, Kutai Basin, Central Sumatra Basin, North Tarakan Basin, and Berau Basin. Explorations started in 2008, and are expected to yield potential development (Wood Mackenzie 2015; CBM Asia 2016; Stevens et al. 2001; Stevens and Hadiyanto 2004).

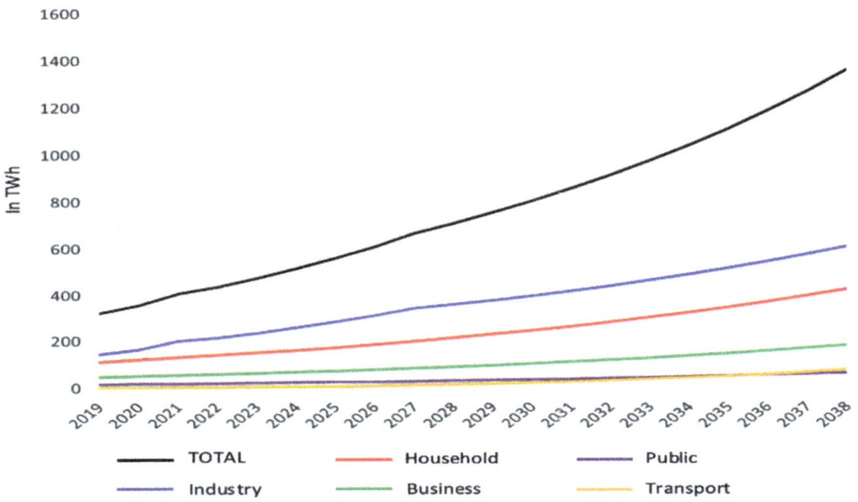

Fig. 2.2 Forecast of Indonesia's electricity needs (2019–2038). (Source: General Plan of National Electrification (2019))

These basins are located in undeveloped areas compared to Java Island with sufficient electricity supply. GoI Constitution Law No. 40/2007 concerning Limited Companies,[4] Article 74, Verse (4), stated that a limited company is obliged to conduct corporate social responsibility (CSR), and that there are penalties for those who failed to conduct CSR.

The two main CBM basins are South Sumatra and East Kalimantan. Both face different challenges. In fact, the challenges between those two basins are contradictory. The local gas market in South Sumatra has been oversupplied by low-cost conventional gas, whilst gas supply in East Kalimantan has been unable to meet the targets for domestic contracted volume and LNG export obligations (Godfrey et al. 2010).

One of the CBM investors, CBM Asia Development Corporation, has a big interest in CBM within Indonesia, interested in the characteristics of coal in Indonesia.

[4] Constitution Law on Limited Company, Ministry of Legal and Human Rights, Republic of Indonesia.

2.6.3 CBM Exploration in Indonesia

Indonesian CBM geology is distinct from all other CBM geologies. Indonesia's CBM basins are located in areas with a dense population. The most important thing to attract prospective investors is the geological prospects and information on geological basins (Chatham House Royal Institutes 2015), and geologically, Indonesia's CBM is attractive. The main potential CBM reserves in Indonesia have met the CBM prospect evaluation criteria. First, Indonesia has thick, abundant, and laterally continuous coal beds. Second, CBM in Indonesia is relatively shallow, with an optimal depth of 1000 to 5000 feet (Dupont and McLelland 2013). The CBM reservoir depth in South Sumatra CBM Basin and Kutai Basin, East Kalimantan, is 2000 feet and 1000–2500 feet, respectively (OffShore Energy 2013). Third, the PRB has proven that low rank of coal can produce more gas and be a successful natural gas project (Harrington 2016). Fourth, Indonesia's basins have favourable reservoirs for the accumulation and preservation of CBM. Fifth, Indonesia's basins have fracturing and faulting evidence in coal. Sixth, Indonesia has a coal-bearing gas-producing field. Seventh, Indonesia evidences good gas content in its main CBM Basins (Saghafi and Hadiyanto 2000; Stevens and Hadiyanto 2004; Davis et al. 2007; Moore et al. 2014).

In discussion with the National Exploration Committee (2015), it was mentioned that CBM exploration and exploitation requires a large working area and a lot of wells to be drilled be economically viable. Such technical and geological challenges cannot be detached from the current fiscal regime. With the current fiscal regime, PSC on CBM, the investors will have to struggle to reach project economy (Bachtiar 2018). Each well that has been tested and proves that it is economically viable needs to get POD approval before it can enter the production stage. Unfortunately, in CBM, only one POD well will not make a project profitable. Similar to Powder River Basin with 1179 CBM-producing wells, Indonesia needs more than a hundred wells to reach project economy (Flores 2004b).

2.6.4 CBM Development in Indonesia

In the development stage of a CBM project, two different wells are required, the water injection well and a production well (Flores 2014b). The CBM production process is unlike conventional gas production process as it requires the dewatering process to create different pressures to allow the gas in the coal to flow to the pipe. However, the climate in Indonesia has been a challenge in the production stage since the climate affects the water content in the ground. This is very much related to the dewatering process in CBM production. The purpose of injecting the water into the coal is to frack the coal and create a path for gas to flow. However, before the gas can flow to the pipe, a different pressure needs to be created via the dewatering process. Wetland environments make the dewatering process a very long one, and sometimes unpredictable in terms of when it will finish and the gas can start being extracted (Husin 2018). Most of the potential basins lie on the equator, with lots of rains and surface water (CBM Asia 2016; Stevens and Hadiyanto 2004), resulting in a wet soil yet extracting the CBM requires water injection. This process can take years, resulting in high set-up costs for investors long before they can produce the gas (Makmun and Sitepu 2012). Unlike Indonesia, the US and Australia have relatively dry climates, and this makes the process easier. As discussed before, CBM exploration and exploitation requires plenty of wells to be economical since the amount of gas produced in one well is relatively low and unstable. In Indonesia, the Plan of Development approval process takes months, and hidden costs will be unavoidable. These hidden costs include the time consumed to prepare documents, waiting time for obtaining the approval, as well as the human resources who prepare for the documents (Saputra 2015). All of these processes delay the investors to start their production stage activities. The PSC also requires the company to be audited for its sunk costs before they get into the development phase. A sunk cost audit is a government procedure to ensure that all the costs incurred are eligible to be recovered through a cost recovery mechanism.

Regulations applied to the PSC for conventional hydrocarbon are also applied to CBM, and these regulations have been a challenge for CBM exploration and development projects (Makmun and Sitepu 2012; Harrington 2016). During the early stage, CBM development requires high capital compared to conventional hydrocarbon (Makmun and Sitepu 2012). Therefore, fiscal incentives are required to support CBM development, and they can take the form of tax relief, capital support, or facilities (Makmun and Sitepu 2012).

2.7 CBM Economy in Indonesia

Currently, Indonesia's CBM projects are at the exploration stage, and there is no CBM development since the fiscal arrangement is not supportive. Therefore, Indonesia needs to have a better fiscal system to support CBM development (Mohamad 2015).

The viability of a CBM project is not only influenced by the gas price. Anderson et al. (2003) wrote that several factors can affect the economics of a CBM project, and that cost efficiency is not the only thing that helps the project economy. The contractor needs to consider technological advances for their drilling, completion, formation evaluation, stimulation, production, as well as reservoir modelling (Anderson et al. 2003). The gas infrastructure and its accessibility have impacts on project viability. Other technical challenges that need to be considered are low initial production rates, high water production rates, and disposal issues. The GoI expected Indonesia's CBM production to reach 500 million standard cubic feet per day (mmscfd) in 2015 (Purwanto et al. 2016; Indonesia Investment 2016; Harrington 2016). In fact, Indonesia only produced 0.625 mmscfd in 2014 (Purwanto et al. 2016; Indonesia Investment 2016; Harrington 2016), and is yet to produce more as many have been terminated (see Chap. 4). Indonesia CBM projects are onshore projects; therefore, CBM investors should consider the population in the area close to them, no small undertaking in such a densely populated country, as evidenced by this situation complicating CBM efforts in terms of water disposal (Cockcroft 2008).

2.8 The Challenges in the CBM Industry in Indonesia

The Indonesian Petroleum Association (2013) stated that unconventional oil and gas (including CBM) in Indonesia were economically marginal projects, but it is worth noting that in 2013 there were no adequate technology, tools, infrastructure, and expertise, or regulations fit for CBM technical purposes. It was still an unproven resource in 2013. The current CBM contractors faced many challenges very much related to PSC implementation. Therefore, two new options on PSC were designed as a response to the problems and challenges faced by the contractors. The first option was the conventional PSC, which is pure PSC; the second option was the Net PSC Sliding Scale proposed by the SKKMigas. The difference was the implementation of Sliding Scale system on the First Tranche Petroleum (FTP) and production split after cost recovery. The third option was a non-cost recovery Gross Split PSC Sliding Scale proposed by the CBM investors.

Net PSC for CBM

The first CBM contract in Indonesia was a Net PSC as applied to conventional hydrocarbons. Gross revenue is based on total lifting multiplied by the standard price, which is being used in Indonesia. The price is the domestic gas price set in President Regulation No.40/2016.

First Tranche Petroleum (FTP) is taken from the gross production before cost recovery. This FTP is like a royalty and a front-end loaded fiscal term (Johnston 1994). The FTP in the current PSC is 20% shareable. Investment credit/uplift rate is an incentive provided by the government to investors to recover its capital expenditures. This investment credit is an additional cost recovered during specific periods, and is taxable (Tordo 2007).

A marginal field incentive is an incentive given by the government to the investors for the working area, which has a rate of return below 15%. This incentive is regulated by the MoEMR Regulation No. 008/2005.

This incentive is an additional cost being recovered amounting to 20% and is a non-taxable item (PriceWaterhouseCoopers 2016).

The MoF regulates the **cost recovery and income taxes** through GR No. 79/2010. MoEMR Regulation No. 22/2008 regulates all other industrial activity costs that are not being regulated in GR No. 79/2010. The contractors can only recover their costs when they start producing. Before this stage, all costs are called sunk costs, and the contractors bear all these costs. The government only shares the exploration risks when they begin to recover their cost through the cost recovery mechanism.

Domestic Market Obligation (DMO) is regulated in Oil and Gas Law 22/2001, GR 35/2004 and GR 79/2010. DMO is an obligation for the contractor to sell oil production to fulfil domestic needs. Based on the regulations, it has to be 25% of the total contractor's equity split. DMO fees may vary depending on what has been agreed in the contract. As an incentive, the GoI usually grants the contractor a DMO holiday, meaning that the contractor sells the production domestically at the market price, commonly only for the first 60 months of the new oil production period. After this, the price used is as agreed in the contract (PriceWaterhouseCoopers 2016). However, CBM production is used to support the government's need to fulfil energy demand (electricity); hence, the DMO holiday is not applicable in this case.

The PSC also applies a **ring-fence policy**, meaning that investors cannot offset the cost from any other exploration field (dryhole field) to a production field (a block that has been granted its Plan of Development). This ring-fence policy gives benefits to the government, enabling them to collect the income tax earlier. This policy is determined to protect government revenue. For a prominent investor with huge potential reserves, this is not a big issue. If the policy is not applied, the investor will recover the costs of the exploration field to the field that is in the production stage. This exploration costs will offset the positive cash flow in the production field and reduce the net income of the investor who will recover the cost amounting to the total cost in some exploration fields added with the total cost in a production field. The government will only get the income tax from a smaller tax base due to the higher cost. The current investors must relinquish 25% of their contract area after six years from signing to the government.

To assist the implementation of the PSC, SKKMigas established the 'Guidelines and Working Procedures' (Pedoman Tata Kerja (PTK)). Unfortunately, the CBM PSC still needs to follow these guidelines regardless of their nature. Seen from its geological condition and technical CBM operational procedures, CBM is more complex compared to conventional hydrocarbons. But there are more challenges than just the fiscal system applied to CBM that have been identified. To illustrate this, the CBM project process prior to starting activities will be discussed.

Oentoeng Suria and Partners (n.d.) identified four challenges in CBM development in Indonesia: (1) overlapping projects in the same field; (2) land clearance and land access; (3) selecting a gas buyer; and (4) financial challenges. Moore (2012) explained the exploration challenges for CBM in Indonesia, mentioning the following three main challenges in developing the industry in Indonesia:

Technical Challenges
Only a few tests had been carried out for permeability, and some of the tests were unreliable. Further, the gas properties varied and required a lot of samples to obtain better data on the gas properties.

Engineering Challenges
Some challenges were identified based on the formation located in Kutai Basin, Barito Basin, and South Sumatra Basin. The challenges in Kutai Basin are very much related to coal permeability and coal thickness. With low permeability, it is necessary to create a better permeability, requiring fracking on thin coal seams. Similar to Kutai Basin, one of the challenges in the CBM industry is that of low permeability. However, the Barito Basin revealed other challenges. Dewatering and water disposal management due to the water content in the coal: the coal in this basin is very wet, and takes a long time for the dewatering process. It also requires reservoir pressure to produce gas. However, the coal rank in this basin is classified as a low-rank reservoir which has a low reservoir pressure. This has been a challenge to produce CBM gas in this basin. The South Sumatra Basin also has dewatering and water disposal challenges for the very same reason, although the coal seams in the South Sumatra Basin are a "co-mingling" of different coal ranks.

Regulatory and Cultural Challenges

As previously mentioned, the same regulations applied to conventional hydrocarbons are applied to CBM. In other words, CBM has been treated as a mature industry, while in fact in 2012 it was very new; it has been in a learning process to develop the industry in Indonesia. It requires time to adopt overseas experiences to the local knowledge in the industry.

With these challenges, the MoEMR established NEC, an ad hoc institution, in 2015 to encourage unconventional hydrocarbon (in general) and CBM (in particular) development in Indonesia. NEC formulated a number of recommendations to drive CBM development in the country.

2.9 Proposed Recommendation for the Development of CBM in Indonesia

Established under the Minister Decree (MoEMR) No. 3097 K/73/MEM/2015 to encourage the exploration activities in Indonesia and aim to increase Reserve Replacement Ratio (RRR) more than 100% in five years by discovering new reserves, NEC is responsible for giving recommendation to the GoI to do their missions. Encouraging unconventional hydrocarbon activities was one of the strategies to increase the RRR.

In 2016, NEC published their recommendations. The following are the proposed recommendations related to unconventional hydrocarbons:

1. The GoI should create special task force involving the Directorate General of Oil and Gas, SKKMigas, and the industry, to construct general guidelines and specific guidelines for new contract term implementation or guidelines for contract transition (old type of fiscal system to new fiscal system) for unconventional hydrocarbon PSC.

2. SKKMigas to establish a separate team for unconventional hydrocarbon exploration activities implementing the Gross Split PSC Sliding Scale as this type of fiscal system is very different compared to the current PSC.
3. There is a need to construct new Work Procedures and Guidelines (PTK) for unconventional hydrocarbon PSC.
4. The GoI to give flexibility in choosing the type of contract available (in 2016) for unconventional hydrocarbons, either the Net PSC or the Gross Split PSC Sliding Scale.
5. The GoI to have a strategy in implementing the Gross Split PSC Sliding Scale and to establish procedures and guidelines for contract transition from Net PSC to Gross Split PSC Sliding Scale.
6. For the Net PSC contract, the NEC proposed some specific recommendation as follows:
    (a) Significant revision on PTK 007;
    (b) POD revision and flexibility entirely become the responsibility of the controller institution;
    (c) Authorisation of Expenditures to be simplified and not applied for each well, as one CBM project requires a hundred wells.

NEC's recommendations were very much related to the new Gross Split PSC Sliding Scale that was approved under Ministry Regulation No. 38/2015, which has now been revoked and replaced by the new Gross Split PSC enacted in 2017.

References

National Exploration Committee. 2015. Jakarta.

Al-Jubori, A., Johnston, S., Boyer, C., Lambert, S., Butos, O., Pashin, J. and Wray, A., 2009. *Coalbed Methane: Clean Energy for the World.* Schlumberger.

Anderson, J., Simpson, M., Basinki, P., Beaton, A., Boyer, C., Bulat, D., Ray, S., Reinheimer, D., Schlachter, G., Colson, L., Olsen, T., John, Z., Khan, R., Low, N., Ryan, B. and Schoderbek, D., 2003. *Producing Natural Gas from Coal.* Autumn 2003. USA: Schlumberger, p. 31.

ASEAN Centre for Energy, 2015. *ASEAN Plan of Actions for Energy Cooperation.* Phase I. [online] Jakarta: ASEAN Centre for Energy, p. 3. Available at: https://policy.asiapacificenergy.org/sites/default/files/HighRes-APAEC%20 2016-2025.pdf [Accessed 9 November 2020].

Asian Development Bank, 2020. *Renewable Energy Tariff and Incentives in Indonesia.* Review and Recommendation. [online] Manila: Asian Development Bank. Available at: https://www.adb.org/sites/default/files/publication/635886/renewable-energy-tariffs-incentives-indonesia.pdf [Accessed 11 November 2020].

Bachtiar, A., 2018. *A Focus Group Discussion on CBM Industry in Indonesia.*

Bryner, G., 2002. *Coalbed Methane Development in the Intermountain West: Primer.* Colorado: University of Colorado Law School.

Calder, J., 2014. *Administering Fiscal Regimes for Extractive Industries.* [ebook] International Monetary Fund. Available at: https://www.elibrary.imf.org/view/IMF071/20884-9781475575170/20884-9781475575170/20884-9781475575170.xml?language=en&redirect=true [Accessed 21 March 2020].

Cassidy, N. and Kosev, M., 2015. Australia and the Global LNG Market. RESERVE BANK OF AUSTRALIA, [online] pp. 33–43. Available at: https://www.rba.gov.au/publications/bulletin/2015/mar/pdf/bu-0315-4.pdf [Accessed 13 July 2021].

Cbmasia.ca. 2016. *CBM ASIA.* [online] Available at: http://www.cbmasia.ca/CBM-In-Indonesia# [Accessed 12 April 2016].

Chatham House the Royal Institute of International Affairs, 2015. *Guidelines for Good Governance in Emerging Oil and Gas Producers.* Energy, Environment and Resources. London: Chatham House The Royal Institute of International Affairs.

Clarke Energy, 2020. *Coal Bed Methane – A Subterranean Energy.* [online] Clarke Energy. Available at: https://www.clarke-energy.com/2011/coal-bed-methane-a-subterranean-energy-treasure/ [Accessed 30 July 2020].

Cockcroft, P., 2008. *Coal Bed Methane in Indonesia.*

Davis, R., Noon, S. and Harrington, J., 2007. The petroleum potential of Tertiary coals from Western Indonesia: Relationship to mire type and sequence stratigraphic setting. *International Journal of Coal Geology,* 70(1–3), pp. 35–52.

Dewan Energi National Indonesia (National Energy Council of Indonesia), 2020. *National Energy Mix 2020.* Jakarta: National Energy Council – Dewan Energi Nasional Indonesia.

Directorate General New and Renewable Energy, 2016. *Buku Statistik EBTKE (Statistical Book of New and Renewable Energy)*. Jakarta: Ministry of Energy and Mineral Resources.

Directorate General of Electricity, 2017. *Electricity Infrastructure Map*. Jakarta: Ministry of Energy and Mineral Resources Republic of Indonesia.

Dupont, P. and McLelland, I., 2013. *Report Analyst of Coal Bed Methane Asia Development Corp*. CBM Asia Development Corp, p. 3.

Flores, R., 2004a. *Coalbed Methane in the Powder River Basin, Wyoming and Montana: An Assessment of the Tertiary-Upper Cretaceous Coalbed Methane Total Petroleum System*. U.S. Geological Survey Digital Data Series DDS–69–C. [online] Denver: U.S. Geological Survey, p. 32. Available at: https://pubs.usgs.gov/dds/dds-069/dds-069-c/REPORTS/Chapter_2.pdf [Accessed 4 November 2016].

Flores, R., 2004b. Wyodak-Anderson Coal Zone in the Powder River Basin, Montana and Wyoming – A tale of uncorrelatable coalbeds. In: W. Miller, ed., *Coalbed methane and the tertiary geology of the Powder River Basin, Wyoming and Montana*. Wyoming: Wyoming Geological Association, pp. 1–24.

Flores, R., 2014a. *Coal and Coal Bed Gas: Fueling the Future*. United States of America: Elsevier Inc., pp. 567–574.

Flores, R., 2014b. *Coal and Coalbed Gas*. Amsterdam: Elsevier, pp. 2–39.

General Plan of National Electrification, 2019. *National Electricity General Plan Republic of Indonesia*. Jakarta: Ministry of Energy and Mineral Resources.

Godfrey, P., Ee, T. and Hewitt, T., 2010. Coal Bed Methane Development in Indonesia: Golden Opportunity or Impossible Dream?. *Journal of Energy & Natural Resources Law*, 28(2), p. 234.

Haldar, S., 2018. Elements of Mining. *Mineral Exploration*, pp. 229–258.

Harrington, J., 2016. CBM Indonesia – Dull Past, Bright Future. In: *2016 Technical Symposium*. Jakarta: Indonesia Petroleum Association.

Husin, M., 2018. *Views on CBM Industry in Indonesia (Investor's Perspective)*.

Ialnazov, D. and Keeley, A., 2020. Motivations, Enabling Factors and Barriers to the Energy Transition in Indonesia and Vietnam. *IOP Conference Series: Earth and Environmental Science*, 505, p. 012044.

IESR, 2019. *Indonesia Clean Energy Outlook: Tracking Progress and Review of Clean Energy Development in Indonesia*. Jakarta: Institute for Essential Services Reform (IESR).

Indonesian Fiscal Policy Office, 2008. *Low Carbon Development Options for Indonesia*. Jakarta: Indonesian Fiscal Policy Office.

Indonesian Petroleum Association, 2013. *Speech on Indonesian Petroleum Association Convention and Exhibition*. Jakarta.

Inter-Governmental Forum – OECD, 2020. *The Hidden Cost of Tax Incentives in Mining*. [online] IGF – OECD. Available at: https://www.oecd.org/tax/beps/hidden-cost-of-tax-incentives-in-mining-draft-toolkit-oecd-igf.pdf [Accessed 20 March 2020].

International Energy Agency, 2014. *Energy Supply Security*. Indonesia. Paris: International Energy Agency, p. 22.

International Energy Agency, 2019. *The Role of Gas in Today's Energy Transitions*. World Energy Outlook Special Report. [online] Paris: International Energy Agency. [Accessed 13 May 2020].

International Monetary Fund, 2015. *"Options for Low Income Countries' Effective and Efficient Use of Tax Incentives for Investment."* [online] International Monetary Fund. Available at: https://www.imf.org/external/np/g20/pdf/101515.pdf [Accessed 20 March 2020].

Investments, I., 2016. *Crude Oil Indonesia – Indonesian Oil Production & Consumption | Indonesia Investments*. [online] Indonesia-investments.com. Available at: http://www.indonesia-investments.com/business/commodities/crude-oil/item267 [Accessed 7 March 2016].

Johnston, D., 1994. *International Petroleum Fiscal Systems and Production Sharing Contracts*. 1st ed. Tulsa: PennWell Publishing Company.

Kemenkeu.go.id. 2020. *Ministry of Finance, Republic of Indonesia*. [online] Available at: https://www.kemenkeu.go.id/publikasi/berita/ini-komitmen-pemerintah-kurangi-emisi-karbon/ [Accessed 30 July 2020].

Kennedy, S., 2018. Indonesia's energy transition and its contradictions: Emerging geographies of energy and finance. *Energy Research & Social Science*, 41, pp. 230–237.

Legowo, E., 2018. *Views on CBM Industry in Indonesia (Government's Perspective)*.

Len, C., 2014. *Indonesia Coalbed Methane*. Bountiful Promise Help Up by Institutional Paralysis. [online] Singapore: Energy Studies Institute. Available at: http://esi.nus.edu.sg/docs/default-source/esi-policy-briefs/indonesia-cbm-bountiful-promise-held-up-by-institutional-paralysis.pdf [Accessed 28 March 2016].

Letts, L., 2012. *Coal Seam Gas Production – Friend Or Foe of Queensland's Water Resources?*. 1st ed. Australia: Thomson Reuters.

Li, S. and Zhang, B., 2016. Research of Coalbed Methane Development Well-Type Optimization Method Based on Unit Technical Cost. *Sustainability*, 8(9), p. 843.

Makmun and Sitepu, E., 2012. Makmun. *Effect of fiscal incentives on coal bed methane price: a hypothetical analysis*, 3(1).

Mallick, N. and Prabu, V., 2017. Energy analysis on Coalbed Methane (CBM) coupled power systems. *Journal of CO_2 Utilization*, 19, pp. 16–27.

Ministry of Energy and Mineral Resources, 2021. Intelligent Strategies Power and Utility Sector to Achieve Indonesia's Carbon Neutral by 2050.

Mohamad, M., 2015. *Progressing CBM Development in Indonesia.*

Moore, T., 2012. Coal Bed Methane: A Review. *International Journal of Coal Geology*, 101(36–81).

Moore, T., Bowe, M. and Nas, C., 2014. High heat flow effects on a coalbed methane reservoir, East Kalimantan (Borneo), Indonesia. *International Journal of Coal Geology*, 131, pp. 7–31.

National Electricity General Plan Republic of Indonesia, 2019. *National Electricity General Plan Republic of Indonesia*. Jakarta: Ministry of Energy and Mineral Resources.

OffShore Energy, 2013. *CBM Asia Updates on Operations in Indonesia – Offshore Energy*. [online] Offshore Energy. Available at: <https://www.offshore-energy.biz/cbm-asia-updates-on-operations-in-indonesia-2/> [Accessed 26 September 2020].

PricewaterhouseCoopers, 2016. *Oil and Gas in Indonesia*. Investment and Taxation Guide, May 2016 – 7th edition. Jakarta: PricewaterhouseCoopers.

Purwanto, W., Muharam, Y., Pratama, Y., Hartono, D., Soedirman, H. and Anindhito, R., 2016. Status and outlook of natural gas industry development in Indonesia. *Journal of Natural Gas Science and Engineering*, 29, pp. 55–65.

Redhead, A., 2018. *Tax Incentives in Mining: Minimising Risks to Revenue.* [ebook] Canada: International Institute for Sustainable Development. Available at: https://www.google.com/url?sa=t&rct=j&q=&esrc=s&source=web&cd=2&cad=rja&uact=8&ved=2ahUKEwjvmofc-cnoAhWpQxUIHQeZDmYQFjABegQIBBAB&url=https%3A%2F%2Fwww.oecd.org%2Ftax%2Fbeps%2Ftax-incentives-in-mining-minimising-risks-to-revenue-oecd-igf.pdf&usg=AOvVaw02chYk5AZ00FBSPJnbvrf3 [Accessed 20 March 2020].

Rosyadi, H., 2010. *Potensi Coalbed Methane Untuk Mensukseskan Program Diversifikasi Energi Nasional.* [ebook] Available at: https://www.researchgate.net/publication/282183107_Potensi_Coalbed_Methane_untuk_Mensukseskan_Program_Diversifikasi_Energi_Nasional [Accessed 22 August 2018].

Saghafi, A. and Hadiyanto, 2000. ABSTRACT: Methane reservoir properties of Indonesian coals. *AAPG Bulletin*, 84.

Saputra, R., 2015. *Preliminary Interview on Plan Development Process.*

Sirait, D., 2013. *Indonesia Current Policy and Regulation.*

Stevens, H., Scott, H., Sani, K. and Hardjosuwiryo, S., 2001. Indonesia's 337 TCF CBM Resource a Low Cost Alternative to Gas, LNG. *Oil and Gas Journal*, (October 22).

Stevens, S. and Hadiyanto, 2004. Indonesia: Coalbed Methane Indicators and Basin Evaluation. In: *SPE Asia Pacific Oil and Gas Conference and Exhibition.* Society of Petroleum Engineers.

Tharakan, P., 2015. *Summary of Indonesia's Energy Sector Assessment.* No. 09, December 2015. Manila: Asian Development Bank, p. 14.

Tordo, S., 2007. *Fiscal Systems for Hydrocarbon, Design Issues.* 1st ed. [ebook] Washington: The World Bank. Available at: https://siteresources.worldbank.org/INTOGMC/Resources/fiscal_systems_for_hydrocarbons.pdf [Accessed 24 April 2016].

United Nations, 2020a. [online] Unfccc.int. Available at: https://unfccc.int/process-and-meetings/the-paris-agreement/the-paris-agreement [Accessed 21 July 2020].

United Nations, 2020b. *All Ndcs.* [online] Www4.unfccc.int. Available at: https://www4.unfccc.int/sites/NDCStaging/Pages/All.aspx [Accessed 9 November 2020].

WoodMackenzie, 2015. *Sumatra CBM.* United Kingdom: Wood Mackenzie.

3

How CBM Projects Align with the Indonesian Energy Transition Goal

Abstract This chapter is about how CBM projects align with the Indonesian Energy Transition Goal. This chapter sets out how CBM projects fit into the goal of increasing clean energy supply in Indonesia with regards to the just transition to a low-carbon economy. Having to tackle climate change, Indonesia has started green energy projects through the development of clean energy supplies such as wind power, solar power, bioenergy, geothermal, and mini hydro- and hydropower. Considering that CBM is a derivative of fossil fuels, Indonesia has developed CBM under the hydrocarbon regimes. It is important to understand that CBM is also known as one of the cleaner fossil fuel sources that can contribute to the additional clean energy supply for the country. The growth of clean energy projects in Indonesia has increased from year to year, and CBM can play an instrumental role in contributing to a cleaner society and environment.

Keywords Coal bed methane • Acceleration • Australia • Barito Basin • Bowen Basin • CBM role • Cost structures • Indonesia • Just energy transition • Kutai Basin • Natural gas • Powder River Basin • Wyoming • Technical challenges • Dewatering management • The US

T. B. Sumarno, *Just Energy Transitions and Coal Bed Methane*, Energy, Climate and the Environment, https://doi.org/10.1007/978-3-030-85490-4_3

3.1 Introduction

This chapter sets out how CBM projects fit into the goal of increasing clean energy supply in Indonesia with regards to the just transition to a low-carbon economy. For example, the benefits of CBM, a gas-replacing coal, is evident in the following quote from the IEA Director: "Natural gas is one of the mainstays of global energy. Where it replaces more polluting fuels, it improves air quality and limits emissions of carbon dioxide. In this analysis, we explore how widespread and durable this role might be in some of today's key energy markets" (Birol, in IEA 2020).

Indonesia is committed to reducing its GHG emissions by 30% by 2030. This chapter provides insights about the role of CBM in achieving these goals. For some countries, CBM has been used for power plant generation. Indonesia has not managed to develop CBM for commercial, nor for power plant generation. Unlike Australia and the US, CBM in Indonesia is more appropriate for power plant generation. Why is this the case? Some geological and technical aspects as well as economic aspects play important roles in the commercialisation of a project, and CBM in Indonesia is currently not commercially viable under the PSC.

This chapter provides a detailed explanation of just how high the cost of a CBM project in Indonesia is in comparison to the US and Australia under different fiscal systems. However, using CBM gas to replace coal in power plant generation could reduce GHG emissions. Not only is CBM gas actually cleaner than coal, it could also drive renewable energy projects to happen faster by having shared facilities (e.g. electricity grid) and, hence, lower their overall costs.

3.2 CBM and Energy Transition Acceleration in Indonesia

There are many strategies to accelerate energy transition that have been discussed by different institutions, such as IEA, IRENA, International Institute for Sustainable Development (IISD), OECD, and many more. One of the strategies is by reallocating fossil fuel subsidies to subsidise the

RE development which eventually lead to phasing out the fossil fuel subsidies. This has been promoted by the IISD and with a support from G20. Another strategy is technology innovation in developing RE, such as *nurture innovation, pursue power-system integration, decarbonise end-use sectors, and expand innovation beyond R&D (such as innovative policy framework)* (IRENA 2017). Dr Fatih Birol stated: "Natural gas is one of the mainstays of global energy: worldwide consumption is rising rapidly and in 2018 gas accounted for almost half of the growth in total global energy demand. Gas plays many different roles in the energy sector and, where it replaces more polluting fuels, it also reduces air pollution and limits emissions of carbon dioxide" (IEA 2020). Natural gas on its own could not be the solution for net-zero emission to tackling climate change. IEA suggested that for emerging economies (such as India and China), switching from more intensive emission energy sources (e.g. coal) to less intensive emission fossil fuel will bring significant reduction CO_2 emission and improve air quality (IEA 2020).

Indonesia, with massive population and as an emerging economy country, should consider to use more gas than other fossil fuels (coal/oil) while also developing its RE. During the Climate Leader Summit in 2021, the president of Republic of Indonesia, Joko Widodo, stated that Indonesia aims to be climate neutral and to have net-zero emission target by 2060. An intelligent strategy was formulated in the energy sector, specifically in the electricity utility sector, which is to replace all of the coal-fired power plant by 2060 (Ministry of Energy and Mineral Resources 2021). However, this strategy includes the 35GW plan by increasing the national coal-fired power plant until 2035. The question is, then, how is Indonesia going to achieve its 29% emission reduction by 2030 while at the same time also increasing the number of coal-fired power plants?

There has been many debates on the energy transition, globally and nationally. Indonesia is proven to have abundant potential resources of CBM. It is important to understand that CBM is not to substitute renewable energy sources in Indonesia, but to add to the energy supply in Indonesia. Indonesia has submitted its Nationally Determined Contributions (NDC) to the United Nations Framework Convention on Climate Change (UNFCCC), that Indonesia has a target of 23% of new and renewable energy in the energy mix and the reduction of coal to 30%

by 2025. CBM is classified as a new energy in Indonesia, and development efforts have been occurring for more than a decade. There have been many challenges in the energy sector in Indonesia, and these include renewable energy development. Steyn (2019) found that CBM impact on air pollution was almost half that of oil and a third of coal. Further, he also looked at the social risk (e.g. health) and societal impacts (e.g. job creation), and he found that CBM as well as natural gas are energy sources with the lowest negative social and environmental impacts compared to oil and coal. Therefore, despite decreases in oil production, Indonesia should consider developing cleaner sources of energy such as CBM.

Indonesia is still at an early stage of its renewable energy development and to meeting its new and renewable energy targets: a 23% target on new and renewable energy by 2025 and at least 31% in 2050 (Government Regulation No. 79, 2014). CBM could support the GoI to reach its target. While it is also important to develop RE in Indonesia, it also becomes significant to have other new sources of energy such as CBM to be developed at the same time.

Each energy project development has its own challenges based on different things, such as geographical location and policy and regulation. Indonesia has learned from the challenges (that will be discussed in the next section) in developing CBM, and this could be an opportunity for Indonesia to develop different sources in addition to RE sources and reduce the usage of high-carbon energy sources such as coal and oil. This will support achieving the ambitions set in the NDC.

3.3 CBM for Electrification Replacing Coal

The main fuel for power plant generation in Indonesia is coal, and developing the CBM in Indonesia will add natural gas into energy mix and replace the coal. The latest example is from Yangquan, China, where a new 30 MW CBM power plant is operational and integrated into the national grid (Yangquan Coalbed Methane power plant 2020). Another example is from Botswana, where they also developed CBM power plants. It is known widely that CBM gas can relatively easily be brought to the surface by drilling wells and extracting it within the coal seam. The gas, which would otherwise be lost into the atmosphere if the coal is mined, is ideally suited for CBM-fired gas turbine plants (Kalahari Energy Botswana 2020). This opens the possibility for Indonesia to develop CBM and use it to electrify some regions in the country. One of big oil and gas companies in Indonesia attempted to develop a CBM project in Sanga-Sanga, Kutai Basin, located in Kalimantan. They entered into an agreement with the National Electricity Company (PLN) to supply 500,000 cubic feet a day (cf/d) over a five-year period (Evans 2011). The first power plant that has CBM as its fuel to generate electricity in Indonesia is Sindang Jaya Power Plant (Evans 2011). However, this was only done once in Indonesia, and there has been no further progress on CBM projects in Indonesia.

The CBM resources in Indonesia are located in Kalimantan, Sulawesi, West Java, and Sumatra Islands. The four largest CBM basins are in Kalimantan and Sumatra. Moreover, the GoI has a commitment to provide full electricity access to all areas in Indonesia by 2020 and reduce GHG emissions by 30% in 2030. CBM in Indonesia is proven to be better used for power plant generation than being commercialised due to its technical and geological aspects and economic aspects. The commercial viability of CBM recovery and utilisation projects is determined by three basic factors: geology, technology, and the economy. Geology and technology determine the technological potential for recovery; the extent of utilisation is largely determined by the economic environment (Thakur et al. 1996). Economic aspects are very much attached to the gas price and cost to extract the gas. In Indonesia, the gas price is regulated under the President Decree No. 40/2016. While the cost of CBM project in Indonesia is mostly influenced by the terms and regulation under the PSC.

3.4 CBM Technical and Geological Aspects

The issue of permeability has been discussed previously. The capability of a CBM well to produce gas can also vary, depending on gas content, saturation, pressure difference, and permeability. The volume and rate of gas production from CBM wells may fluctuate significantly, unlike conventional gas, which is often more consistent once it starts producing. According to Williams et al. (2001), minimum or low gas CBM producers yield about 50,000 cubic feet (mcf) per day; high-yield wells produce as much as 5 mmcf per day. Therefore, as previously mentioned, one CBM project could require a hundred wells in order to be economically viable in delivering a certain volume of gas, unlike conventional natural gas.

3.5 Economic Aspect 1: Gas Price

Selling gas is more difficult than selling oil. To be able to sell gas, the producer is required to initially find the buyer and build the infrastructure, which costs four times more than for oil (Stickley 2014). According to Stickley (2014), there are four elements for the gas industry to be successful: first, it requires a large amount of gas to be commercially viable; second, it requires infrastructure such as pipeline systems; third, the gas market has a price constraint which is happening in Indonesia, and the gas price is regulated by the government; fourth, ways by which the government can provide the incentives for the activities. According to Golubchenko (2018), a petroleum fiscal analyst from WoodMackenzie, it is not easy to find a buyer:

> even for operators with bigger amount of gas, they still find that gas allocation, or finding buyers is very hard. (Golubchenko 2018)

Project commerciality is determined by the revenue and cost of a project. Revenue is very much attached to the volume and price of the commodity. However, in a gas project, the volume is defined in the gas sales agreement, but the price is either using the market price or the regulated price. In Indonesia, the gas price is regulated by the MoEMR, including CBM gas. The CBM cost was initially very high due to the PSC and its attached procedures and regulations, and, lately, the CBM investors have managed to reduce the costs as they are now able to use proper technology for CBM operations (Husin 2018). Boosting CBM commerciality is done by increasing the gas price, but it will become a challenge for the investors as it will make the CBM gas less competitive compared to conventional gas.

> The next problem is the CBM commerciality. CBM commerciality is not easy. However we will sell our gas from CBM, but the buyer may not have the same confidence level as they have when they buy the gas from conventional hydrocarbon. Since it has not been proven yet. CBM is a marginal project, and we are not looking for profit, but its margin, and added with the market that is not sure with its potential. This makes the commerciality even more difficult. [...] we expect that the Government understand and then can help this industry to compete with the conventional hydrocarbon, that's what we expect. [...] Even if we manage to save the industry, it does not mean we can commercialise the project. (Setyatmoko 2018a)

> in the beginning of 2009/2010 we used mining rig to push the cost down, yet the cost was still very high. I don't understand why. It may be caused by the mobilisation cost. [...] So, even there has been an innovation for CBM technic, which is one of the key thing, but in the reality, it was still not enough to push the production cost down [...] yes that was because the gas price. If only the gas price was still $ 14, there will be many investors jump into CBM. (Husin 2018)

NuEnergy Ltd. has a concern regarding the gas market and the competitiveness of CBM gas in the market. At this moment, the CBM gas is not very competitive comparing to the conventional gas, as it is challenging to create a competitive gas price for CBM gas. According to Mr. Setyatmoko, since the gas price is regulated, and unless the government is aware and takes action on this, the market will never be competitive for CBM gas. Another challenge is that the CBM gas market is not yet stable, unlike conventional gas. People are more aware of conventional gas than CBM gas (Setyatmoko 2018b). Regardless of all the challenges faced by them, NuEnergy Ltd. is the only CBM investor that is moving forward. It is believed that when they prove that their CBM is commerciality viable, there will be an opportunity to revive the industry in Indonesia.

CBM gas, then, tends to be sold and used domestically, mainly due to commerciality issues. Another reason is due to regulation. According to

Golubchenko (2018), it is easier to sell gas domestically because it is more feasible as the market is easy to reach. However, in Indonesia, it is more difficult to sell gas domestically because of the limitations, which are regulation and politics (Golubchenko 2018). Yet it has been more than ten years since the first CBM contract was signed (Purba 2018):

> But if you are gonna go for domestic gas supply, then you go for the domestic gas price, you can price whatever it works. Whatever it works to reach for that village or, for that power plant or factory then whatever. Then it would work. But there is gonna be flexibility there that you can just make it work for all parties. And that's bit tricky. Because then you have power politics and crisis. (Mair 2018)

> The price will have to compensate for the fiscal system. So you can either say reduce government take or you can increase price. (hahaha) it is just leading the problem to downstream. (Mair 2018)

The investors and government may set the gas price at any price that works for them to reach the set goals of that village or other parties, but flexibility is still required to make things work for all parties (Mair 2018).

3.6 Economic Aspects 2: CBM Cost Structures

Coal in each basin has different physical, chemical, and compositions (Rightmire 1984), leading to different methodologies of producing the gas in each basin. As previously discussed, coal seam depth and coal permeability determine the cost of CBM projects. This project cost is also linked to the cost of water production (Meredith et al. 2010; U.S. Department of Energy 2007). CBM production cost in Indonesia is strongly related to the cost of water disposal and the cost of each CBM well.

CBM costs in Indonesia were estimated by the GoI (BPMigas 2010a), yet this cost estimation is far from the actual CBM cost. There are many factors that the GoI failed to anticipate in this estimation, and the actual CBM cost is closer to that estimated by the existing CBM investors. The exploration cost of CBM includes geological and geophysical studies, land acquisition, exploration drilling costs, and general costs related to this activity. Acquisition of a large area of land is essential for CBM projects due to the number of wells required in one CBM project (ALL Consulting 2004).

3.6.1 Comparative Study on CBM Cost Structures

One of the main CBM basins in Indonesia, Kutai Basin, is located in East Kalimantan. There have been many CBM activities in this area, and all the PSC commitments have been fulfilled by the investors (SKKMigas 2018). As such, this location makes a useful case study.[1] Moreover, the analysis shows why the PSC fails to support CBM growth in Indonesia from a non-technical point of view. The data was analysed quantitatively by calculating the project economy with PSC, and qualitatively by analysing non-budgetary items, such as the regulations and work procedures carried in PSC implementation. Likewise, a streamlined comparative analysis of CBM cost structures in Indonesia, the US (Powder River Basin), and Australia (Bowen Basin) was also conducted for comparative purposes.

Data and Assumptions

Kutai Basin

The Kutai Basin has similar coal characteristics with the coal in PRB (Stevens and Hadiyanto 2004). There have been CBM activities in this basin since 2009, and it has a successful water disposal management in place, and shared facilities from the existing conventional gas infrastructure (Balfas 2012; Munawar et al. 2013). The depth of the source rock in the Kutai Basin is 1600 to 3300 feet, and the well tends to be shallower than those of a conventional well (Stevens and Hadiyanto 2004).

Cost structures consist of capital cost and operating costs. Tangible costs in the development stage are classified as capital costs, as well as the

[1] Based on the Author's research in 2017–2020.

water disposal facilities costs. The intangible costs in the development stage are classified as operating costs. Field facility costs are broken down into capital costs and operating costs. All general and administrative costs related to the activities are classified as operating costs.

Data on estimated costs and production were obtained from focus group discussion among the GoI and CBM investors.[2]

Additional Assumptions

- The project is estimated to obtain POD after seven years of exploration stage and start its production at year 8.
- The total project life is 30 years, meaning the production field life is 23 years after entering into the production stage.
- This project is aimed to be extended for another 20 years.
- Each production well is estimated to have 20 years of production life; its production rate can reach up to 200mscfd, and the declining rate is 12%.
- The total production during the project life is expected to be 1.5 TCF by drilling 2300 wells, including core hole and pilot well. The peak production rate for each well is expected to be 200 mscfd for three years.
- Prices used are the based price (US$ 6/mcf), as defined above, low price (US$ 4/mcf), and high price (US$ 8/mcf), according to 2% escalation. Whilst the costs used are the based case (US$ 3MM/development well), low cost (US$ 2MM/development well), and high price (US$ 4MM/development well). There is 2% escalation on prices and no escalation on development costs (CAPEX), as it is assumed that the costs have been determined in the contract for building the facilities. Decommissioning and abandonment costs are given.
- The NPV is calculated at 10% discount rate with the following consideration:

---

[2] Some Focus Group Discussion that were held involved data specific in related to CBM project in Indonesia.

- The average annual general inflation rate in Indonesia for ten years (2007 to 2016) is 6.22% (Bank Indonesia Inflation Data [accessed in February 2020])
- Buffers (reserves) of -/+ 3% which are generally used by SKKMigas. This is to anticipate national inflationary movements due to external factors and the challenges of risks stemming from global uncertainty and the monetary policies of developed countries.
- There is no DMO applied to this fiscal regime since the production supports domestic needs.
- The production incurred during the exploration phase is shared between the government and contractor using the model being applied.
- All regulations and work procedures applied for conventional hydrocarbon are applied to this project.

Powder River Basin[3]

Data was taken from a 2006 US Department of Energy study, written by Bank and Kuuskara (Advanced Resources International Inc. 2006). All financial data has been adjusted to 2010. The fiscal regime applied to CBM in PRB is the royalty and tax fiscal regime.

Bowen Basin[4]

Data was taken from Massarotto (1999). All financial data was in AUD, and it was 1997 financial data. The Australian data is limited, and the current data is not publicly available. Hence, this data is adjusted to USD. The fiscal regime applied to CBM in the BB was the Petroleum Resources Rent Tax regime prior to the Treasury Amendment in July 2019.

[3] There has been limited data access on commercial data for PRB and the Bowen Basin.

[4] Ibid.

3.6.2 Comparative Analysis of CBM Cost Structures

This section provides a comparative study of CBM project cost structures in the US (Powder River Basin), Australia (Bowen Basin), and Indonesia (Kutai Basin). The purpose of this comparative study is to investigate the fundamental causes of the cost differences and discover the reason why PSC in Indonesia fails to support CBM growth.

Kutai Basin (KB)

The cost of a CBM well in the Kutai Basin appears to be the most expensive of the three basins examined here. The comparable cost is the estimated cost by SKKMigas (previously BPMigas) in 2010 except for well costs and the exploration and development cost per well. These costs are the adjusted costs, which was the investors' estimated cost after adjustment. Since the estimated costs are not detailed, the data from SKKMigas for the water gathering cost, water disposal (treatment), electrical power, and gas gathering is used. Land and permit costs are determined by the government. The cost for well drilling and completion is much higher than those in the PRB and BB.

Powder River Basin (PRB)

The report by Advanced Resources International Inc. (2002) found that water management was one of the most important considerations impacting on the costs and the feasibility of the projects. Very strict water management practice does not support the development of the industry, eventually leading to the loss of recoverable CBM resources, reduced governmental income (royalty, state severance, and ad valorem), and also reducing the capital investment. Therefore, improvement in water management is significant to maintain project viability.

The report also discussed the basic economic model for PRB CBM. Some basic costs in this economic model are as follows:

- Capital Costs, which includes land, permits, drilling, and completion (D&C), and infrastructure. The depth and the location of the wells determine the capital costs.
- Operating and Maintenance costs for the wells, which depend on the year of the production.
- Gas distribution and transportation, which depends on the charges for transporting the gas which also depends on its location. This is also called as infrastructure costs.
- Water disposal alternatives, which have impact on the project costs.
- Other things that are considered as company's costs are all the costs paid to the government, such as royalty, tax, and/or state severance.

Bank and Kuuskraa (2006) evaluated the CBM project economy with a base gas price scenario of US$ 4/mcf. At this price, the field development capital per CBM well is $ 191,400. For each dollar in increased price, the cost per well drilling and completion went up by 22%. The cost of well D&C at US$ 4/mcf gas price is US$ 93,000. These capital costs include land costs and permits, drilling and completion, infrastructure and water management costs. The capital expenditure per development well are much influenced by the depth of the coal sources. The coal seams depth in PRB range from 500 to 950 feet (Bank and Kuuskraa 2006). The water disposal method used here is surface discharge, known as the cheapest method of water disposal management. The cost of water disposal being presented above is the capitalised water disposal costs. The variable cost of water disposal being categorised as operating costs, which are US$ 0.04/bbl.

Bowen Basin (BB)

The cost of a CBM well in Australia tends to be more expensive than in the PRB. Cost of leasing and acquisition in Australia is comparably more expensive, reaching up to US$ 800,000. In general, the CBM cost in Australia is 25% higher than the CBM cost in PRB (Massarotto 1999).

The first CBM exploration in Australia was in BB in 1970, while the first CBM extraction occurred in central Queensland in 1996 (ICF International 2014). In the BB, the coal seams depth ranges from 2300 to 3300 feet.

Table 3.1 shows the CBM costs comparison of the three basins[5] based on the existing data. It shows that the CBM cost items in Indonesia are the highest, especially well drilling costs, which were eight times higher than the US and Australia. There was no data on water disposal costs of the CBM in the US and Australia and no data on electric power and cable costs for CBM in Indonesia. Therefore, these are not comparable.

There are various reasons why the costs are different, particularly in Indonesia. These include technical and non-technical reasons that eventually influence the technical challenges. The technical reasons are very much related to the geology and natural condition of the CBM in Indonesia, as well as the technical expertise on CBM; the non-technical reasons are related to the government's fiscal policy applied to the CBM industry.

Table 3.1 CBM cost structures in Powder River Basin (per well)

Cost items	Indonesia	Australia	The US
Land cost and permits	US$ 120,000	US$ 35,100	US$ 16,200
Well drilling and completion	US$ 2,500,000	US$ 105,300	US$ 89,200
Water gathering	US$ 300,000	No available data	US$ 27,700
Water disposal	US$ 250,000	No available data	US$ 1600
Electric power including cable	No available data	US$ 21,100	US$ 13,400
Gas gathering	US$ 100,000	US$ 70,200	US$ 47,500

Source: NN (2015); BPMigas (2010a); Massarotto (1999); Bank and Kuuskraa (2006)

[5] The capital cost of PRB CBM well (Bank and Kuuskraa 2006) at US$6/mcf with an inflation rate of 8.2% (US Inflation Calculator). The capital cost of Bowen CBM well is adjusted by applying the inflation rates at the annual inflation rate of 2.8%, from 1997 to 2010 (Reserve Bank of Australia 2017). These adjusted costs are then being converted into US$ using the exchange rate of 31 December 2010 (XE currency exchange).

1. **Technical reasons**

(a) *Coal seam depth*
The coal seams depth in PRB ranges from 500 to 950 feet. This is considered shallow compared to the depth in BB or Indonesia, which ranges from 2300 to 3300 feet and 1600 to 3300 feet, respectively (Massarotto 1999; Stevens and Hadiyanto 2004; Davis et al. 2007; Moore et al. 2014). Ideally, the shallower it is, the cheaper is the cost. However, in Australia, the cost is lower compared to Indonesia. This is explained in the next sections.

(b) *Level of expertise*
The existence of CBM potential in PRB was initially identified in 1957, and an official survey from the US Geological Survey was first done in 1978 by drilling 15 holes and testing gas content (Flores 2004a). At around the same time (the 1970s), Australia started its CBM project in BB (Thurber and Morse 2015). The CBM activities in both basins have been ongoing for more than 40 years. Meanwhile, in Indonesia, CBM potential was first identified in early 2000, and the first CBM project was started in 2008. The CBM industry in Indonesia is relatively new compared to the other basins. As a beginner, Indonesia is still in its learning curve to cultivate a domestic CBM industry.

(c) *Existence of technology*
Technology is associated with the level of expertise. By 2010, both the US and Australia have been learning for more than 40 years and have developed better technologies for the industry. Indonesia has limited technology experience[6] (Setyatmoko 2018b). Before 2016, the investors were required to procure rigs for conventional gas. There are procedures to procure the particular drilling rigs for CBM that the investors are obliged to comply with, creating barriers for investors.

(d) *Water quality and climate*
Locations and climate play a significant role in determining the CBM project cost. Indonesia is a tropical country with a high precipitation

[6] The first CBM block was awarded in 2008 (SKKMigas 2016).

level, and it rains a lot more compared to the U.S. and Australia. This is linked to the water management disposal system. There are five water management practices being examined (Bank and Kuuskraa 2006) with different levels of costs:

Surface discharge
: This has the lowest cost amongst all. Produced water is pumped from several wells to one central location, and the water is discharged directly into a river. The produced water must be safe, and the quality must meet government standards. This is not the case for Indonesia since the CBM produced water may contain substances that can be harmful to the environment.

Infiltration impoundments
: This is the second-lowest cost after surface discharge. Produced water is pumped and gathered into the centre of a pond, and the water is evaporated. This has been applied to both water disposal in PRB and BB. This option is possible in Indonesia, but since Indonesia has high precipitation, it will be difficult. Furthermore, due to the rain, there will be a huge amount of water, requiring a bigger pond.

Shallow reinjection
: More expensive than both surface discharge and infiltration impoundments since it requires both pond and reinjection wells to dispose of the water after the produced water is processed and filtered, VICO CBM chose this option to dispose of their produced water and utilise their unused well as a reinjection well to reduce costs (Munawar et al. 2013).

Reverse osmosis
: This option is the second-highest cost amongst all, involving the development of more facilities, including water-holding and storage impoundments for residual substance, and water treatment system for reverse osmosis. TDS discharge limits constrain the amount of produced water being disposed of. Reverse osmosis costs can be up to US$ 1.5 million.

Ion exchange
: The most expensive water treatment system, and similar to reverse osmosis, this option requires more complex facilities and water treatment for ion exchange which can cost up to US$ 2 million.

2. **Non-technical reason: The type of the fiscal regime**
   Both the US and Australia implement a royalty/tax system for their CBM industry, allowing them to manage and control their expenditure. The Australian government offered fiscal incentives to CBM investors at their beginning of CBM development in Australia, such as a royalty holiday and a 15% investment tax credit. These incentives increased the IRR to 1.6% (Massarotto 1999). In Indonesia, the PSC system requires investors to comply with the guidelines to allow them to recover their expenditures (cost recovery mechanism). The drilling cost in Indonesia was six times higher than the US and three times higher than Australia. This was due to the relatively higher cost of the drilling rig in Indonesia since they had to use the rig for conventional gas (IPA 2016). This system and procedures also cost due to the longer time required to obtain all kind of approvals to run the project.

3.6.3 Indonesia's CBM Cost Analysis

The depth of the CBM source rock in Kutai Basin is 1600 to 3300 feet, and the CBM well tends to be shallower than conventional wells. However, since CBM wells may produce gas up to 5 mmcf/day, it requires more wells to drill. A CBM project has a lower risk at the exploration stage compared to conventional hydrocarbons (Godfrey et al. 2010). Higher risks in CBM projects are incurred in the appraisal, development and production phases, resulting in higher costs. Unlike conventional gas, the CBM capital costs also occur outwith exploration phase, including the project phase due to the production cycle and volume. Gas production increases in the early stage along with a decrease in water production. Figure 3.1 shows the production profile of a CBM project and its capital expenditures. The production volume for each CBM well is low; therefore, the drilling of more wells are required.

The PSC brings both technical and financial implications. There are specific requirements in the guidelines and work procedures that must be complied with, such as local content requirement, tender process requirement, accounting procedures until the end of the project, and

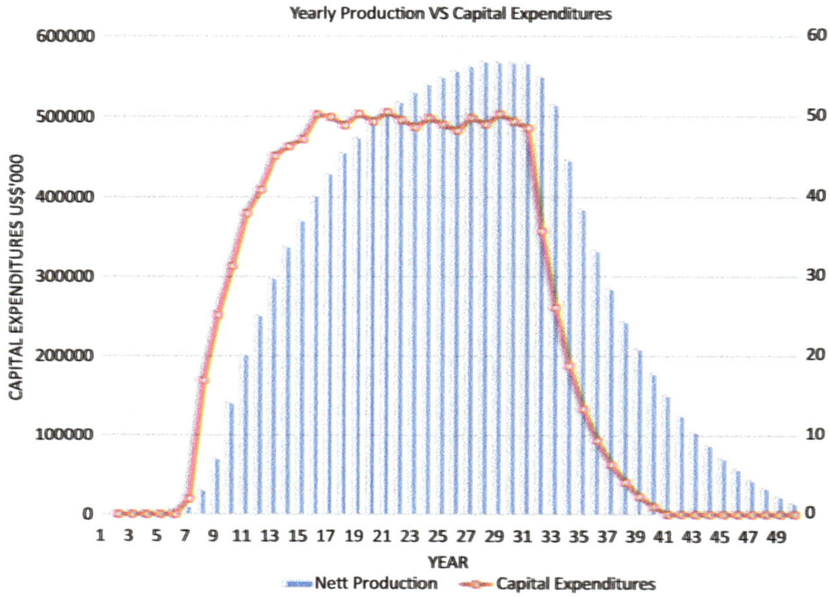

Fig. 3.1 Yearly CBM production versus capital expenditures (Source: Created by the author (2020))

abandonment and site restoration procedures. This compliance is crucial as it will impact the costs being recovered through the cost recovery mechanism.

Using the data assumption presented earlier, the exploration costs incurred in the first six years following contract signing amounted to US$ 71,400 M. These costs consisted of G&A and field operating expenditures (OPEX) during the exploration stage and the core hole and pilot well costs. Assuming one core hole each year from year 2 to year 5, and the drilling of six pilot wells each year in year 4 and year 5, the total costs amounted to US$ 50,4MM. All of these programs and expenditures had been proposed in the WP&B and approved by the SKKMigas likewise for the related AFE. Each goods and services procurement must comply with the PTK 007 (Guidelines and work procedure for Goods & Services Procurement); therefore from the initial action to finding goods/services providers, the investors are mandated to comply with the rules.

Considering all of these processes, the investors must spend extra time to be able to run the exploration activities. As long as there is no production lifting, these costs are considered as sunk costs and can only be recovered once they start producing gas from the production wells. After all, the investors have taken on the sole risk of the exploration activities.

The investors can only start their production activities after they obtain approval of their Plan of Development proposal for a CBM working area. Unlike POD conventional gas, the POD for CBM is determined by working area instead of by structures. This is a government initiative to support production activities since many wells are required fpr CBM to be economically viable. This POD must include content on the geological findings and reviews, reservoir descriptions, reserves and production forecasts, drilling and completion, production facilities, field development scenarios, health, safety, and environment, CSR, ASR plans, project schedules and organisation, local content, and commerciality (PTK 072, 2010). Although the CBM POD does not carry financial costs, it does take a one to two month period for the GoI to assess the POD proposal, in the case that everything goes well. Such complexity just to reach the production stage creates problems for investors.

Assuming that POD approval is obtained in year 7, the investors can then start production in year 8. As previously discussed, CBM capital costs are incurred throughout the lifetime of the CBM production cycle.

In the PSC system, the capital cost is recovered through a depreciation mechanism, regulated by the guidelines and work procedures for accounting (PTK 059) set by SKKMigas. This mechanism delays the investors ability to cope with their expenditure. There is no cap for a cost recovery item. However, it has 10% First Tranche Petroleum. This FTP is actually the limit of the cost incurred to be recovered in that year. The total cost that can be recovered in year A is as much as the total gross production after FTP, meaning only 90% of total gross production. If the total cost in year A is more than 90% gross production, the rest of the cost is carried forward to the next year.

The gas produced during the exploration stage is shared between the GoI and the investor based on the agreed split. However, the cost recovery mechanism only occurs after POD approval, as shown in Fig. 3.2.

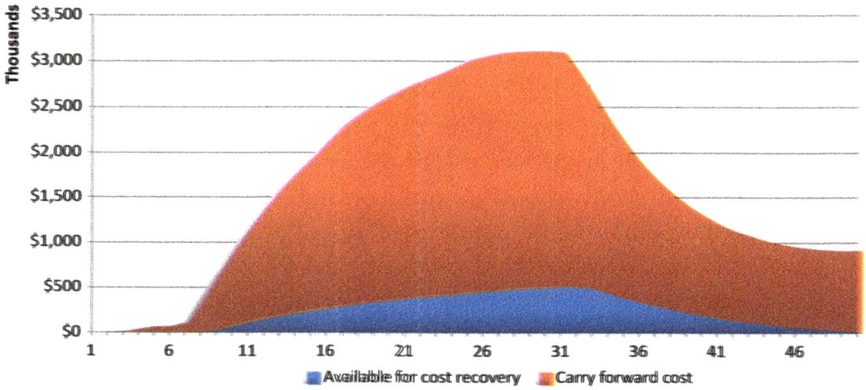

Fig. 3.2 Available for cost recovery VS unrecovered cost (carry forward cost). (Source: Created by the author (2020))

Thus, it is very unlikely for the investor to recover their costs at the end of the project under the current PSC.

This PTK also provides guidance on imposing the cost in the context of ring-fencing policy. It is challenging to apply the same ring-fence policy applied to conventional projects for CBM projects due to the number of CBM wells being drilled. It requires plenty of time and administrative work to prepare a POD proposal for each well that is going to be developed. The POD's PTK for CBM is currently being developed.

Located in East Kalimantan, Kutai Basin is in an area with a high precipitation rate since it is on the equator. According to the PSC, the investor of the CBM project in this basin is allowed to utilise the existing facilities from a nearby conventional project (Munawar et al. 2013). However, the CBM project requires separate water treatment facilities due to the produced water quality and volume uncertainties (Munawar et al. 2013). The volume of water produced in one field of a CBM project can vary.

Although water surface disposal is the cheapest method, it is unlikely to be applied in this basin due to the climate. The CBM project in this basin applies two water management methods, impoundment and reinjection (infiltration impoundments), which have led to higher costs. As such, the idle wells from other PSC's blocks are used as the point of water injection to reduce costs. It is important that the water disposal is purified in the pond before being reinjected into the well (Munawar et al. 2013). This water disposal and management is regulated under Regulation No. 04/2007 of the Ministry of Environment and Forestry.

What Other Things Create a High Cost in CBM?

That Indonesia is an archipelago country has created another problem for transfer, and connectivity between the islands is one of the challenges. This challenge has become more complex since inadequate infrastructure is in place to ease the transfer. However, since CBM is planned to be locally used, distribution will not be a problem. Therefore, this issue is more related to material, tools, and equipment mobilisation:

> the challenge in Indonesia is the infrastructure. There is no infrastructure availability which distinct this with the US. the infrastructure availability. The second one is in Indonesia, why is it difficult? It is because the connectivity, we are different with the mainland. [...] Indonesia is an archipelago country. (Lubiantara 2018)

The equipment standard in the Net PSC applied to conventional oil and gas is very costly, yet the CBM project is required to follow the standard, regardless of differences between conventional and unconventional hydrocarbon production:

> it's about the equipment. The equipment we are using from the early stage, drilling stage till the pumping, it's all with conventional oil and gas standard, and it is very expensive. (Rahmadhi 2018)

According to Gunawan (2018), the GoI has changed its requirement for procuring the CBM tools and equipment, through learning from past experience about CBM and having the willingness to adjust requirements to support the industry. It has now become cheaper, and the investors should take advantage of this situation. However, the initial costs incurred in the early stage of CBM development involved such a massive amount of cash that they are now facing financial difficulties.

References

Acim.nidec.com. 2020. *Yangquan Coalbed Methane Power Plant.* [online] Available at: https://acim.nidec.com/generators/leroy-somer/references/yangquan-cbm-power-plant [Accessed 30 July 2020].

Advanced Resources International, Inc., 2006. *The Economics of Powder River Basin Coalbed Methane Development.* Prepared for: U.S. Department of Energy. [online] Advanced Resources International, Inc. Available at: http://www.adv-res-hou.com/pdf/The%20Economics%20of%20Powder%20River%20Basin%20Coalbed%20Methane%20Development.pdf [Accessed 27 September 2017].

ALL Consulting, 2004. *Coal Bed Methane Primer.* New Source of Natural Gas – Environmental Implications. Montana: U.S. Department of Energy, National Petroleum Technology Office.

Balfas, R., 2012. Successful CBM Coring Activity Integrated with Conventional Well Drilling in Sanga-sanga Block, Mahakam Delta, East Kalimantan, Indonesia. In: *Indonesian Petroleum Association.* Jakarta: Indonesian Petroleum Association.

Bank, G., and Kuuskraa, V., 2006. *The Economics of Powder River Basin Coalbed Methane Development.* Washington D.C.: Advanced Resources International, Inc.

BPMigas, 2010a. *CBM Cost Estimation in Indonesia.* Jakarta: BPMigas.

BPMigas, 2010b. *Type of POD and Approval Body.* Jakarta: BPMigas.

Davis, R., Noon, S. and Harrington, J., 2007. The petroleum potential of Tertiary coals from Western Indonesia: Relationship to mire type and sequence stratigraphic setting. *International Journal of Coal Geology*, 70(1–3), pp. 35–52.

Evans, D., 2011. *Indonesia's CBM Dream Nears Reality.* [online] Petroleum-economist.com. Available at: https://www.petroleum-economist.com/articles/upstream/exploration-production/2011/indonesia-s-cbm-dream-nears-reality [Accessed 30 July 2020].

Flores, R., 2004a. *Coalbed Methane in the Powder River Basin, Wyoming and Montana: An Assessment of the Tertiary-Upper Cretaceous Coalbed Methane Total Petroleum System.* U.S. Geological Survey Digital Data Series DDS–69–C. [online] Denver: U.S. Geological Survey, p. 32. Available at: https://pubs.usgs.gov/dds/dds-069/dds-069-c/REPORTS/Chapter_2.pdf [Accessed 4 November 2016].

Flores, R., 2004b. Wyodak-Anderson Coal Zone in the Powder River Basin, Montana and Wyoming – A tale of uncorrelatable coalbeds. In: W. Miller, ed., *Coalbed methane and the tertiary geology of the Powder River Basin, Wyoming and Montana*. Wyoming: Wyoming Geological Association, pp. 1–24.

Godfrey, P., Ee, T. and Hewitt, T., 2010. Coal Bed Methane Development in Indonesia: Golden Opportunity or Impossible Dream?. *Journal of Energy & Natural Resources Law*, 28(2), p. 234.

Golubchenko, N., 2018. *Views on CBM Industry in Indonesia (Global Consultancy's Perspective).*

Gunawan, B., 2018. *Views on CBM Industry in Indonesia (Government's Perspective).*

Husin, M., 2018. *Views on CBM Industry in Indonesia (Investor's Perspective).*

ICF International, 2014. *Recent Australian Natural Gas Pricing Dynamics and Implications for the U.S. LNG Export Debate*. ICF Resources.

Indonesian Petroleum Association, 2016. *Oil and Gas Sector to Be the Motor of The Economy.*

International Energy Agency, 2020. *The Role of Gas in Today's Energy Transitions.* [online] Paris: International Energy Agency. Available at: https://webstore.iea.org/download/direct/2819?fileName=TheRoleofGas.pdf [Accessed 10 November 2020].

IRENA, 2017. Accelerating the Energy Transition through Innovation. WORKING PAPER based on global REmap analysis. [online] Abu Dhabi: IRENA. Available at: https://irena.org/publications/2017/Jun/Accelerating-the-Energy-Transition-through-Innovation [Accessed 13 July 2021].

Kalahari Energy Botswana, 2020. *How Is Coal Bed Methane Used – Kalahari Energy Botswana.* [online] Kalahari Energy Botswana. Available at: https://keb.co.bw/how-is-cbm-used/ [Accessed 30 July 2020].

Lubiantara, B., 2018. *Views on CBM Industry in Indonesia (Local Consultancy's & Government's Perspective).*

Mair, D., 2018. *Views on CBM Industry in Indonesia (Global Consultancy's Perspective).*

Massarotto, P., 1999. Cost Benefit Analysis of Coalbed Methane Recovery Activities in Australia and New Zealand- Implications for Commercial Projects and Government Policy. In: M. Mastalerz, M. Glikson and S. Golding, ed., *Coalbed Methane: Scientific, Environmental and Economic Evaluation*. Dordrecht: Kluwer Academic Publishers, pp. 33–54.

Meredith, E., Wheaton, J., Kuzara, S., Donato, T., Bierbach, S. and Schwarz, C., 2010. *2009 Annual Coalbed Methane Regional Ground Water Monitoring Report: Northern Portion of the Powder River Basin, Montana*. Open File Report 591. Montana: Montana Bureau of Mines and Energy.

Ministry of Energy and Mineral Resources, 2021. Intelligent Strategies Power and Utility Sector to Achieve Indonesia's Carbon Neutral by 2050.

Moore, T., Bowe, M. and Nas, C., 2014. High heat flow effects on a coalbed methane reservoir, East Kalimantan (Borneo), Indonesia. *International Journal of Coal Geology*, 131, pp. 7–31.

Munawar, A., Hendromurti, D., Dewi, R., Foglio, G. and Schupp, B., 2013. Dewatering Facilities for CBM Resource Appraisal, Lesson Learnt from Sanga-sanga CoalBed Methane, East Kalimantan. In: *Indonesian Petroleum Association*. Jakarta: Indonesian Petroleum Association.

NN, 2015. *CBM Project Economics Calculation*. [Microsoft Excel] Jakarta.

Purba, S., 2018. *Views on CBM Industry in Indonesia (Government Perspective, Written Interview)*.

Rahmadhi, H., 2018. *A Focus Group Discussion on CBM Industry in Indonesia*.

Reserve Bank of Australia. 2017. *Inflation Calculator | RBA*. [online] Available at: https://www.rba.gov.au/calculator/ [Accessed 6 May 2019].

Rightmire, C., 1984. *Coalbed Methane Resources of the United States*. Tulsa: American Association of Petroleum Geologists, pp. 1–13.

Setyatmoko, U., 2018a. *A Focus Group Discussion on CBM Industry in Indonesia*.

Setyatmoko, U., 2018b. *Views on CBM Industry in Indonesia (Investor's Perspective)*.

SKKMigas, 2018. *Coal Bed Methane Blocks in Indonesia 2008–2018*. Jakarta: SKKMigas.

Skkmigas.go.id. 2016. *SKKMIGAS » Statistics of Production*. [online] Available at: http://www.skkmigas.go.id/en/statistik/statistik-produksi [Accessed 4 April 2016].

Stevens, S. and Hadiyanto, 2004. Indonesia: Coalbed Methane Indicators and Basin Evaluation. In: *SPE Asia Pacific Oil and Gas Conference and Exhibition*. Society of Petroleum Engineers.

Steyn, J. 2019. *Hydraulic Fracturing of Rock Formations Part 2: Coal-bed Methane Recovery*. 1st ed. [ebook] Owner Team Consultation. Available at: <https://www.researchgate.net/publication/336441808_Hydraulic_Fracturing_of_Rock_Formations_Part_2_Coal-bed_Methane_Recovery> [Accessed 26 September 2020].

Stickley, D., 2014. *A Framework for Negotiating and Managing Gas Industry Contracts*. 2nd ed. Dundee.

Thakur, P., Little, H. and Karis, W., 1996. Global coalbed methane recovery and use. *Energy Conversion and Management*, 37(6–8), pp. 789–794.

Thurber, M. and Morse, R., 2015. *The Global Coal Market*. Cambridge: Cambridge University Press.

U.S. Department of Energy, 2007. *Produced Water Management and Beneficial Use*. 2nd Semi-annual progress report. Colorado: Colorado Energy Research Institute.

Williams, R., Yurakov, E. and Ashelford, D., 2001. Gas Emissions Modelling of Gate Road Development. In: *Coal Operators' Geo-technology Colloquium*.

4

Current Issues on CBM Development in Indonesia: What Do They Say?

Abstract The next chapter reveals the real issues regarding the development of CBM projects in Indonesia from different stakeholder perspectives including investors, government, and consultants. Categories and classification of issues are developed and explained in this chapter. It also provides an understanding that there are different ways of seeing a project, yet these issues are interrelated and complex. Furthermore, issues arise due to the impact(s) of other issues; therefore, solving issues in isolation will not push CBM development forward. To identify and solve these issues, this chapter will look at two developed countries, the US and Australia, and how they have had successful CBM development and benefited hugely from these projects (Hulme, *Coalbed Methane Development in Wyoming's Powder River Basin.* Colorado: University of Colorado, 2002; Bryner, *Coalbed Methane Development in the Intermountain West: Primer.* Colorado: University of Colorado Law School, 2002; Mackinnon and Fox, *Wyoming Law Review*, 6(2), pp. 369–400, 2006; Mudd, *Golden Gate University Environmental Law Journal*, 5(2), pp. 297–342, 2012; Letts, *Coal Seam Gas Production – Friend or Foe of Queensland's Water Resources?* Australia: Thomson Reuters, 2012; Cully, *Review of the Socioeconomic Impacts of Coal Seam Gas in Queensland.* Canberra:

T. B. Sumarno, *Just Energy Transitions and Coal Bed Methane*, Energy, Climate and the Environment, https://doi.org/10.1007/978-3-030-85490-4_4

Australian Government, Department of Industry, Innovation, and Science, 2015).

Keywords Coal bed methane • CBM challenges • National paradigm • Regulation • Fiscal issues • Government • Local community • Fiscal policy • Commerciality

4.1 Introduction

This chapter sets out the current issues facing the development of CBM projects in Indonesia from different stakeholder perspectives: investors, government, and consultancies.[1] Different categories and classification of issues are developed and explained in this chapter. It also builds on the understanding that there are a number of ways of viewing a project and how it develops, and that these issues are interrelated and complex. To identify and solve these issues, this chapter examines CBM projects in two developed countries that have had successful CBM development and benefited hugely from doing so. The two countries are the US and Australia (Hulme 2002; Bryner 2002; Mackinnon and Fox 2006; Mudd 2012; Letts 2012; Cully 2015).

4.2 The CBM Current Situation in Indonesia

The data used in this chapter is very diverse. NuEnergy Ltd. is the only CBM operator in Indonesia currently being reviewed by SKKMigas for its Plan of Development proposal, under the Net PSC. As such, the challenges faced by the company are very different from those faced by other CBM investors. Another company is currently attempting to avoid being penalised as their blocks are being terminated by the GoI due to their failure to fulfil the firm commitments. The idea is to transfer their commitment from the unfavourable block to a block known for its positive

[1] This date was obtained from interviews that the author held with various stakeholders involved in CBM development in Indonesia. See also Sect. 4.3 for a more detailed discussion of this.

prospects. A similar situation is faced by another domestic CBM investor who are currently facing financial difficulties due to incurring massive expenditure in their initial project.

The GoI, investors, and consultancies (as well as the industry association) all contributed their thoughts about how they saw the current CBM industry in Indonesia. They have a different way of seeing the industry at this moment. The government agreed that the current investors whose blocks are in the process of being terminated focused more on avoiding the penalty. However, they also understand that Indonesia still retains the opportunity to develop the CBM industry. This is despite the consultancies showing that the industry is slowing down. There have been many issues that have hampered the growth of CBM in Indonesia.

According to the investors' interviews, the preliminary results from the field CBM study (from 2002 to 2004) showed that the start of CBM development in Indonesia was inadequate. The investors admitted that they were impulsive and overexcited to initiate the development. As of 2008, 7 CBM blocks were awarded, rising to 54 CBM blocks by 2015.

According to SKKMigas (2018), a total of 24 CBM blocks were terminated by 2015. Up to 2017, six CBM blocks were forced to terminate by the government, with investors proposing the termination of another five with the condition that the firm commitments were fulfilled. The main CBM investor proposed that another nine blocks be given approval to transfer their commitments. However, not a single one of the nine blocks received approval, and indeed eight of them were, in fact, being terminated in 2018, with the last one under the termination process. Forced termination is mainly due to the failure of the investors in fulfilling their firm commitments over a given period, namely the first three years in their exploration period, according to the PSC contract (Government Regulation No. 34/2005, Article 8). It is crucial for investors to have these firm commitments when bidding for blocks, as regulated under Ministerial Regulation (MoEMR) No. 5/2012 on the Procedures for Determining and Tendering Unconventional Working Contract Areas.

This has brought investors who are under the termination process into a situation where they need to put more effort into avoiding their financial obligation due to the failure of fulfilling their firm commitments. Known as a penalty, this financial obligation is the total financial amount of their remaining unfulfilled firm commitments deducted by the value of their Performance Bond (Government Regulation No. 30/2017):

> Penalty on termination for investors who fail to fulfil their firm commitments is imposed on (a) Contractor that has proposed to relinquish all their working contract area and has been approved by the Minister to be terminated; (b) Contractor that has been recommended by the SKKMigas and has obtained approval from the Minister to be terminated; or (c) Contractor that the contract is expired automatically and has obtained approval from the Minister to be terminated. (MR No. 30/2017, Article 5, Clause 3)

> The amount of the penalty is calculated by the SKKMigas based on the residual value of the firm commitments that are failed to be fulfilled by the contractors. (MR No. 30/2017, Article 10, and Clause 2)

> The payment of the penalty on contract termination as stated in clause (2), is using the residual value of the Performance bonds and/or cash. (MR No. 30/2017, Article 10, and Clause 3)

Currently, these investors are negotiating through SKKMigas to get a penalty relief from the government by proposing to transfer the commitments instead. The idea of transferring the commitments is being discussed between the government and the investors via SKKMigas:

> the mission of the CBM investors at the moment is to survive and free themselves from liabilities, penalty. (Gunawan 2018)

> our focus currently is to save the blocks under termination process so that we don't get the penalties. (Rahmadhi 2018)

The discussion concerning transfer commitments is expected to ultimately encourage investors to fully fulfil their commitments. The current PSC contract does not allow the investors to transfer commitments, yet it requires the government to support it in terms of regulation. However, Gunawan (2018) stated that it was easier to amend the contract than create a new government regulation considering the time required and the complexity involved:

> currently, we are discussing transfer commitments. So, the firm commitments in the field with less good property or low prospect, that can be transferred to another field which … this is still being discussed. And yes, needs to be supported by the regulation. (Gunawan 2018)

> (transfer commitment) according to the contract, it breaches the contract. A reference in a contract cannot be transferred just like that to another contract. […] therefore, now, we are in the process of creating the Government Regulation, This is still debated as it does not mean the regulation can change the contract. […] so yes, it is easier to amend the contract. (Gunawan 2018)

A completely different idea was raised during the interviews, that of Underground Coal Gasification (UCG). The possibility of developing a UCG project was proposed by one of the main CBM investors for some of their non-potential CBM projects. According to Gunawan (2018), however, this project would incur the same problems as CBM, namely to do with the technology and the costs:

> we have proposed to swap our commitments to SKKMigas, and it was also attended by the Directorate of General Oil and Gas from the MoEMR, and they didn't agree with that. […] I have heard from. Gunawan, that the issue was not in the SKKMigas but in the Ministry of Finance. So what happened was that from the joint study result, it was found that they have potential on CBM, but then when we explored them, they are not potential at all, for example at Kuala Kapuas Block. So, we proposed to swap the commitments, but maybe, I don't know what the SKKMigas consideration was, I meant, they were not brave enough to go to that direction. And the last thing is, how we supposed to get the gas out of this block. But with the gas content which was not as much as we expected before, permeability, we all know that these no longer can be expected. Last time, we thought of proposing this coal potential for UCG to SKKMigas. (Rahmadhi 2018)

Despite all the difficulties and current challenges faced by the investors, together with the GoI, they have learnt more about the technical aspects of CBM. As a result, in 2016 the government allowed investors to use mining rigs. This decision has made the industry more cost-efficient as there is no daily rate when using a mining rig, and it is a lump-sum base cost. Further, the mining rig service company takes the risk if there is delay in the process:

> at the end of 2016, we used the mining rig. [...] there is no daily rate for mining rig, but they are lump sum. (Husin 2018)

The GoI is aware and acknowledges that Indonesia has some areas that have CBM potential. They have classified areas with CBM potential into areas with high CBM prospects, medium CBM prospects, and low CBM prospects. The GoI has started showing their support to the industry by increasing the availability of the tools and equipment, including the rig for CBM. This has brought down initial costs by two-thirds:

> now we have a view, which area has the highest prospectivity, which area has medium prospectivity and which area with the lowest prospectivity. (Gunawan 2018)

> lately, the procedures are the same, but there is availability for tools and equipment. CBM rigs, added by ... I meant the tools and equipment supply, there are more and can be cheaper. Last time it decreased to USD$600,000, so it was reduced to one third from the initial cost. (Gunawan 2018)

However, these projects have been stalled due to many reasons. The government and the investors were not quite ready at that time, resulting in the stalled projects. Today, many more projects are stalled. The number of active CBM block has decreased to 43 at the beginning of 2018, and by the end of 2018 it dropped significantly to 30. The data shows that there were three more blocks that are under the termination process in 2018 (SKKMigas 2018).

Fig. 4.1 The number of CBM projects in Indonesia (2008–2018). (Source: Created by the author (2020—data collected from SKKMigas 2018))

In summary, 1 block was terminated in 2015, 5 blocks in 2016, another 5 blocks in 2017, and 13 blocks in 2018. Moreover, there were still 3 CBM blocks currently under the termination process as they failed to fulfil their commitments (see Fig. 4.1).

There has been no further CBM project development since the first CBM contract was signed ten years ago. Many big investors have left Indonesia, and those remaining are small investors who keep trying to fix the situation. There has been no proven project since the first seven CBM blocks were awarded. However, the Indonesian CBM industry may grow if the government can provide adequate support for the industry to prove that at least one project is profitable. According to Pujobroto (2018), there will only be one or two that can:

> the current situation is now dying. So if the PSC – net PSC works, only one or two can actually be said to have a future in CBM. The rest, I think they will just trying to fulfil their commitment.

The CBM industry has faced many challenges, yet the government has taken considerable time to fix the regulation that in the end could not even be implemented. The new fiscal regime, Gross Split PSC, officially implemented in January 2017, remains to be properly implemented due to the incomplete regulation and taxation regulation (for the Gross Split PSC). Furthermore, the implementation, when it happens, will bring more challenges for the industry. Currently, some CBM investors consider that the Gross Split PSC is better than the Net PSC (Wajong 2018b; Setyatmoko 2018b; Husin 2018). The concept of Gross Split PSC initially came from the CBM investors through the IPA, the industry association. In reality, this will add more complexity for those current CBM investors that have implemented the Net PSC. One of the biggest challenges is the sunk cost that they have spent during the exploration stage. This is another significant issue facing the industry:

> to finally get the new system approved took 2 years at least. [...] The ministry regulation has been set but could not be implemented since it is still incomplete. The taxation regulation to accompany the new system is still not clear, and the SKKMigas has not very accepted the new system in place. (Husin 2018)

> the current situation, nobody is interested, and not even the government is interested, and it's for the remaining operators, it's getting hard to justify spending more money on CBM. Because there is absolutely no support. (Golubchenko 2018)

> big CBM players who had high expectation, left the country, since they saw this project was not economically prospective. And the rest of us, we have tried to make the cost more efficient, and then improve the exploration, the exploration data. [...] But you know what, eventually, the situation is not as we had expected and then we try to fix it, but it has been too long, it's been too long the process to fix the problem, and at that time, the regulation was in placed yet was properly implemented, but yes, it has been too long. (Setyatmoko 2018b)

> Let the data talk. How long has been since the first CBM contract was signed. Has any of them already achieved to Commercial status. You may

> see one or two along and around or in adjacent with VICO field. But have you ever found any single CBM Contract that already pass the stage of commerciality? (Purba 2018)

In a sector characterised by little movement, NuEnergy is the only CBM investor moving forward. It is believed that when NuEnergy proves CBM commerciality, there will be an opportunity to revive the industry in Indonesia:

> I think it is basically on the last breath, I think other than NuEnergy I am not seeing mm any operator that are still moving forward. Mmm but I haven't been in touch with mm who is the biggest equity holder? CBM holders? Sugico, yak, I haven't been in touch with Sugico. And I would still say that with Pertamina as hearing back as they can, I think NuEnergy Ltd. is the only one that is moving forward and seeing results from that. Yeah, and probably it would be if NuEnergy can pull something off, prove that they can achieve what their plan of development field to do, then probably there is a chance to revive the industry, but again, even at best,, what I am expecting is, at best probably for smaller project like NuEnergy Ltd, you know, just somewhere at most 20mmscfd/project. (Utama and Golubchenko 2018)

These situations have been triggered by issues that arose from different perspectives provided in the interviews. These have been classified into five categories:

1. Government-related issues
   The issues classified in this category are those not necessarily related directly to the fiscal aspect, but are directly related to the government's decisions, reactions, responses, and involvement in the industry.
2. Fiscal issues
   These issues are directly related to the fiscal aspect, and bring impacts regarding project commerciality.
3. Local issues
   The issues very much related to the local communities and environment in which the project is sited/operated.
4. Investor-related issues

The issues specifically related to the investors' current situations, which are not influenced by the government's decision or the fiscal environment, including financial capability and human skills.

5. Technical issues

   The issues classified in this category are related directly to CBM technology and geology. Not necessarily influenced by the fiscal or regulatory actions set by the GoI, nonetheless it is essential to understand and address the technical issues to be able to move the industry forward.

4.3 Interview and Focus Group Discussion: The Challenges in the CBM Industry in Indonesia

The oil and gas industry has faced many issues as a whole, and the CBM industry in Indonesia is no exception. Lubiantara (2017) classified these into four main issues: the instability of oil and gas law and regulation; conflict of interest amongst governments and sectoral bodies; government paradigm; and industrial risk, high costs and limited budgets.

This book reveals similar issues in some areas but some more specific to the CBM industry in Indonesia. These are classified into five issues: government-related issues, fiscal issues, local community–related issues, investor-related issues, and technical issues. Figure 4.2 shows the classification of the issues created by the author as a result of the interview, the FGD and the survey analysis. These are explained in detail in the subsequent sections of this chapter.

4.3.1 Government-Related Issues

The biggest concern regarding issues related to the GoI is the lack of policy formulation for CBM, which is seen as ignorance on the part of the GoI (Bachtiar 2018). It was mentioned during the FGD that the GoI has not focused on the energy industry, but rather on national infrastructure. This ignorance has worsened the energy situation in general, with

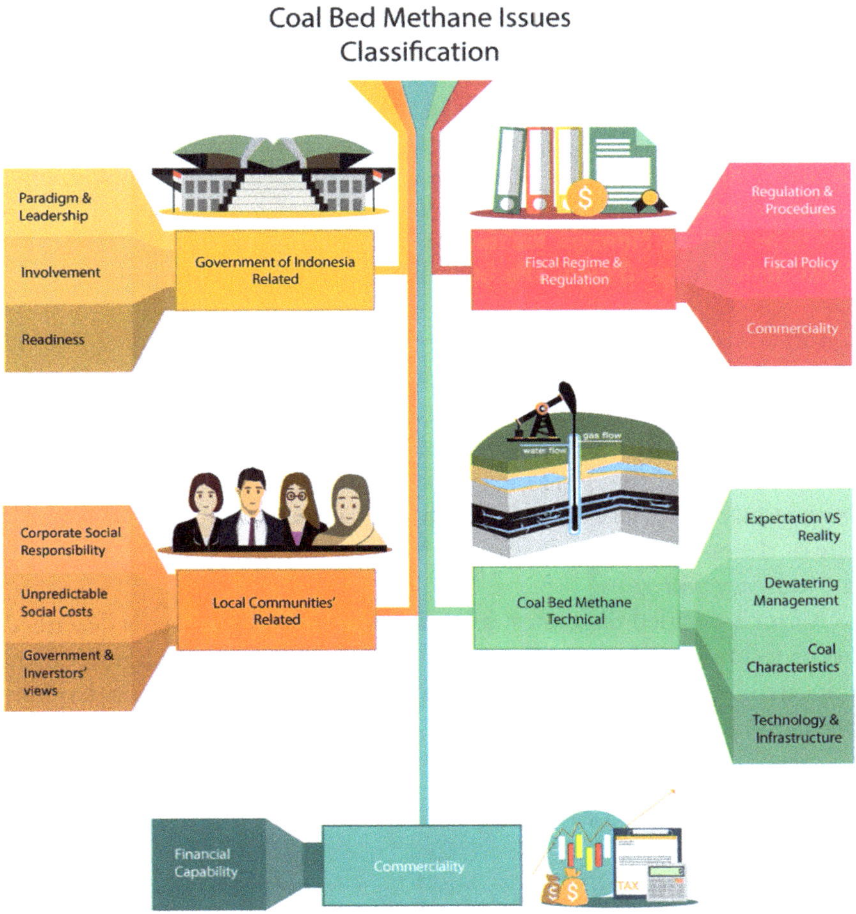

Fig. 4.2 Issues and challenges in the CBM development in Indonesia. (Source: Created by the author (2020))

particularly negative consequences for the CBM industry. All related issues return to the ignorance of the GoI.

According to Bachtiar (2018), the GoI has been ignorant of the domestic CBM industry. For example, the energy mix target had been discussed since 2007/2008 and was finalised in 2014 by enactment of Government Regulation No. 79/2014 (NEP):

> One of the articles states about the energy mix target, and in that context, it was not mentioned how much they should be getting from the CBM. There is nothing mentioned at all. CBM is mentioned, but no number at all. In the energy policy, we look at the energy mix, what those are, how much we expect from each of the energy. This has been started since 2007/2008, prepared in 2009 and just finalised in 2014, and CBM is not mentioned there as in how much is being expected for the energy mix; as well as nuclear. However for the rest of the energy types are stated, how much they targeted them. (Bachtiar 2018)

Ignorance towards the CBM industry is real, and the government has no clear expectations for the sector. According to Bachtiar (2018), the government should have included a target for CBM in the NEP.

Legowo (2018), the former director of General Oil and Gas in 2008–2012, said that the government should care more about CBM in Indonesia. She believes that when the government starts to see CBM as one of the energy resources that can bring significant benefits for society, the industry will grow.

A much-detailed discussion related to the GoI's ignorance is described in the following subsections, centred around the GoI's paradigm, leadership, involvement in energy activities, and readiness in terms of developing CBM in Indonesia.

The Paradigm and Leadership

Indonesia has substantially improved its infrastructure, with the exception of the energy industry. According to Bachtiar, if the leader of a country commands the people to improve the energy sector, it will definitely improve. Critically, this would depend on how and where the leader motivated the people to move into.

A new paradigm on energy resources was framed by the National Energy Council (NEC) of the Republic of Indonesia in 2014. NEC has four missions: (1) to design and frame the NEP of the country; (2) to define the countermeasures of crisis and emergency energy conditions; (3) to oversee implementation of the cross-sectoral energy policy; and (4) to make the NEC an effective and trusted independent institution.

The government enacted Government Regulation (GR) No. 79/2014 to realise the NEP framed by the NEC: "The energy resources are no longer as export commodity only, but as the engine of national growth" (Article 6, GR79/2014). Moreover, energy resources are enable a prosperous society as much as possible by optimising its benefits for national economic development (Article 7, GR 79/2014).

As part of its mission, the NEC reviewed the implementation of the energy policy, attached in President Regulation No. 22/2017, which explains the General Energy National Plan (RUEN). RUEN is the mandate of Energy Law No. 30/2007, established according to the NEP.

RUEN underlined one of the main issues in the energy industry, which is the change of the paradigm. The government has not treated the energy resources as the engine of growth, instead viewing it as a source of revenue. There is no alignment on the derivative rules, procedures and regulations with the primary energy laws, for example the Oil and Gas Law No. 22 2001 stated that new investors could acquire permits from only three different government institutions: the Directorate General of Oil and Gas Republic of Indonesia, SKKMigas, and the Department of Finance under the Ministry of Finance Republic of Indonesia. This statement contradicted with the fact that investors are still required to obtain different permits from different institutions including the Ministry of Forestry for land access and Local Government before they can start their activities (Tambunan, n.d).

The missed implementation of the paradigm was also mentioned many times during the FGD and individual interviews. Bachtiar, as a member of NEC at that time, mentioned that the government never actually implement the paradigm, and only put it formally in the regulation:

> The Government's approach is still the same as the old approach, which was if we can make money out of it, let's make it all. If we can tax up front, let's do that. If we can make revenue from it, even for the data, we can get money from it, etc. let's do it for the Government's source of revenue. (Bachtiar 2018)

Setyatmoko, the current CBM investor, agreed with Bachtiar. However this does not only apply to CBM industry, but the energy in general, and as it applies to CBM, it has become worse since the CBM is a very cost-intensive project, yet has not been proven.

> the key is energy as the engine of growth, and not for the source of revenue. The pattern of the CBM industry is for the source of revenue and not as the engine of growth. And this is not just for CBM but for all energy types. [...] yet the government paradigm is still as the source of revenues. (Bachtiar 2018)

The GoI wanted their shares even before the investors produced the gas from the CBM. This showed that the government's paradigm never changed from "energy as a commodity" to "energy as the engine of growth" (Soeryowibowo 2018).

According to Lubiantara (2017), there exist four national paradigms to establish energy security in Indonesia:

- Pancasila as the country's ideology
  The oil and gas sector must be based on the Pancasila, meaning that the industry shall give benefit to the welfare of society. CBM should be used for the greatest benefits of society. Therefore, this research looks at the local community in which CBM development incurs.
- 1945 Constitution
  The 1945 Constitution forms the background of the PSC in Indonesia. The resources underneath the country's land are to be used for the greatest benefits of society. As previously mentioned, the aim is to use CBM gas for local society (to enhance electricity supply).
- National archipelago insights
  This paradigm is strongly related to how people see their country and their environment. The GoI should pay more attention to how the industry, including the CBM industry, can support economic growth.
- National security
  CBM should become one of the commodity reserves for national security. According to the former director of General Oil and Gas, Legowo, the CBM should be one of Indonesia energy resources used for the benefit of society.

Indonesia is one of the most naturally resource-rich countries in the world, and according to Gylfason and Zoega (2006), a country with an abundance of natural resources tends to experience a slow down their economic growth. Indonesia has fallen into the resource curse (Gylfason and Zoega 2006). Gylfason and Zoega (2006) stated that there were four routes for natural resource-rich countries to slow down economic growth. In this case, Indonesia has fallen into rent-seeking behaviour, for example, imposing tariff protection for domestic producers. Indonesian society, led to believe that their country is rich, tend to have a false sense of security. Further, the GoI neglect the need of having good and growth-friendly economic management, such as bureaucratic efficiency and institutional quality. It seems, therefore, that the GoI failed to infuse their paradigm into their policy and actions.

Involvement

Government involvement is one of the most critical factors for successful CBM development. There are many ways that government can be involved in CBM apart from setting regulations and rules relevant to the CBM industry.

The CBM prospect in Indonesia is quite promising, but there must be a direct involvement by the GoI into the sector/projects, especially in the medium-to-high prospective areas. Once CBM commerciality is proven, it will invariably attract more investors. The author identified seven key points where the government should be more involved: (1) the key performance indicator (KPI), (2) responses, (3) land access, (4) local participating interest, (5) regulated gas price, (6) affordability, and (7) internal management issues.

Involvement 1: Key Performance Indicator (KPI)

The government should think comprehensively and not see the industry from one point of view without considering the other important factors that influence the growth of the CBM sector. According to Soeryowibowo

(2018), the GoI currently does not think comprehensively to support the industry's growth. For example, on setting the KPI, they are not aligned, and some are contradictive, for example between the Ministry of Finance (MoF) and the Ministry of Energy and Mineral Resources (MoEMR):

> Since they don't think comprehensively, the way they think is spotting. Do you know why? Each ministry has KPI, correct? Each KPI in each ministry is different and very contradictive. And as a result, they are not coordinated one and another. And even worse, they are conflicting. (Soeryowibowo 2018)

In the Attachment (the Guidelines) of the Decree of the Ministry of Finance, No. 467/KMK.01/2014, on Performance Management in their Ministry, it is clearly stated in Chapter II that one of the KPI for the Directorate of General Taxation in the ministerial level is the government's revenue, which includes total tax revenues, total customs revenues, and total non-tax revenues. These are targeted in the APBN. This KPI has crossed the MoEMR's KPI, which is related to government revenue. Prior to GR 79/2014, one of the MoEMR KPIs is to meet the APBN for its revenue. However, in 2014, as previously discussed, the GoI set a new paradigm concerning energy resources, namely that they should be seen as the engine of growth rather than simply a source of revenue. Having adopted this paradigm, the GoI should have focused more on indirect economic growth as one of the impacts of energy development. Natural resources have been one of the country's main revenue sources. However, such a focus has ultimately led to an unattractive system considering that the investors hold the exploration risks and that their interest is to create maximum profits for their shareholders. This can be seen by looking at the ring-fence policy that does not allow the investors to transfer their exploration cost from a dry hole field to their production field. This ring-fence policy deters investors from exploring new fields (frontier field). They have high chances of failure in finding the gas (exploration stage), yet the ring-fence policy means that they are unable to transfer the cost to the production field. From the investors' perspectives, removing the ring-fence policy will give them the ability to transfer this cost and to reduce their tax payable to the GoI as it will offset revenue from the production field. In reality, the GoI reduces their income tax revenue if they fail to apply the ring-fence policy. Having the ring-fence policy, then, does benefit the GoI via higher taxes, as any dry hole expenses are born by the investors. However, the reality is that the GoI failed to implement their own paradigm, as they are still focused on revenue growth.

Each government body/department has a different vision, missions, and work programmes that they must achieve, and often in the process of achieving these, they conflict with other departments, thus hampering meeting national goals (Lubiantara 2017).

The Involvement 2: Responses

The GoI must be more realistic about the energy resources in Indonesia (Legowo 2018). The current challenge for the GoI is to design an attractive PSC or fiscal regime for unconventional hydrocarbons, especially CBM (Soeryowibowo 2018).

In January 2017, the GoI enacted a new fiscal regime for the oil and gas industry, which also applied to the CBM industry. Called the Gross Split PSC, this was considered better for CBM by some correspondees, such as Legowo, Wajong, Setyatmoko, Rahmadhi, and Husin (2018).

However, the GoI has not been supportive of the investors since the implementation of MR No. 33/2006 on CBM Exploration and Exploitation Activities (Noor 2015). Regardless of the implementation of their regulation on CBM Acceleration Activities (MR No. 38/2015), in reality, little effort has apparently been made to ease the switch from Net PSC to the Gross Split PSC (Wajong 2018b). For example, if they wish to switch their fiscal regime, some ongoing processes (e.g. POD) will be delayed (Wajong 2018b). Wajong also stated that there was no real action from the GoI in implementing the new regime (Utama and Golubchenko 2018; Legowo 2018).

Another example is that when the government finally developed a strategy on CBM activities via MR No. 38/2015, this was revoked even before the investors could gain from the regulation.[2] One of the CBM companies under POD process approval would have switched to Gross

[2] This was replaced by MR No.08/2017.

Split PSC for their CBM block if the GoI had been quicker in responding to the issue (Setyatmoko 2018b).

Setyatmoko (2018b) also stated that another issue was that investors had limited time to wait for a government response on the taxation system. Investors have commitments to fulfil, and uncertainty provided by the government has forced them to keep going with the Net PSC regime, despite not being that economically viable for them (Setyatmoko 2018b).

The Involvement 3: Land Access

The government can also get involved in land access matters. According to Bachriadi and Suryana (2016), formal land use change is the initial stage and fundamental for the issuance of business operation and land use permits. Prior to starting the CBM project, the investors must acquire the land and get access to the land. This is one of the common challenges in the oil and gas project, especially for CBM development project. In the CBM project, the investors drill more than just a hundred wells;[3] therefore, it requires a lot of land, and most of the CBM resources in Indonesia are located either in densely populated areas, in a palm oil plantation areas, or coal mining areas. Each of these has different challenges for CBM investors. Issues with the local people often happen in the oil and gas areas, in this case CBM areas. It is agreed by interviews participants (2018) that the GoI should guarantee that the investors can do their activities without any unnecessary disturbances from local/other parties. This could be an opportunity for CBM investors to involve local communities and drive the economy in the area. This of course requires governmental support, and assistance will be crucial for CBM investors:

> the challenge is that it requires us to get the land acquisition, most importantly land acquisition. The locals see this as their opportunity that will not come twice in their thought, or very rare. For us, we must be careful in spending our money, cos we do not know yet, if we will get something, yet they want the money right away. So that is the challenge which can be a challenge in any other industry and not just CBM. Therefore, the

[3] The number of wells for a CBM project is very much related to the characteristic of the CBM itself.

> Government should deal with the land acquisition process, otherwise everyone will claim about the fairness of the industry. So yes, the Government should see this and make it fair. (Wajong 2018b)

> The Government shall clear all the licenses and guarantee the investors to be able to start their operation right away. That's all. Would the Government do that? ...[...] The thing is, does the GoI have the willingness to do that? It is again not necessarily the government is willing or not, but rather, are there people who can deal with this? (Soetrisno 2018)

> apart from the local people, who else are there? Palm company. That is what make us discsouraged from investing. (Soetrisno 2018)

> if you want to open an area for your CBM activities (say) in Sumatra, then you start your process, and suddenly, there are banana trees showing up in your area. Suddenly there are sheds there,. [...] and then you will get complaints from the local people, saying that you must pay for the cracks on their wall – which it may not be caused of your activities, and it has been there even before you started your project. So this is what discouraged us at that time. (Soetrisno 2018)

Gunawan said that there was a discussion on land acquisition with the vice minister about the formalities, particularly for land acquisition. The idea was to get the land permit done collectively by SKKMigas, as mentioned in the Law on Land Agrarian, No. 2/ 2012 Article 6 and Article 52:

> In the Land Acquisition being done for infrastructure to support oil, gas and geothermal activities, is procured based on strategic plan and Institution's work plan who requires land [...]. (Article 6)

> In the case an institution requires land, Government Legal Institution / State-owned Enterprise that is appointed, funded by the internal company or other sources according to the regulation. (Article 52)

The government's legal institution for the oil and gas industry was BPMigas, which is currently SKKMigas:

> Lately, when there was an event by the Vice Minister at SKKMigas, we all gathered. What if for the formalities, most importantly the formalities, for the land acquisition – surely by the contractor- but for the permits, what if we do it collectively by the SKKMigas. There is regulation for this, but I forgot, where exactly, maybe in the agraria law or what. [...] so now, the SKKMigas has begun to assist the land acquisition process and those permits approval for both conventional and unconventional hydrocarbon. (Gunawan 2018)

The Involvement 4: Local Participating Interest

The government imposed 10% of the local participating interest to the company, as stated in Government Regulation (PP) no. 35/2004 Article 34:

> Since obtaining the POD Approval for the first field that is going to be developed in a Contract Area, the Contractor is obliged to offer the 10% participating interest to the Local Government's institution.

The idea of imposing the local participating interest was to ease the process of starting the project in the local area. The intention of imposing this interest is good, yet the GoI has failed to implement it properly (Husin 2018). The local participating interest was meant to create a sense of belonging to the project for the local people, so that the local community could maximise its benefits from having an energy project in their area. This has become a challenge due to the derivative regulation for the implementation on the 10% local participating interest only being enacted in 2016 by Ministerial Regulation No. 37/2016. The investors did not consider carrying the 10% sunk cost, which was supposed to be borne by the local government. The regulation obliged investors to offer the 10% to the local government institution appointed by the government (MR No. 37/2016 Article 9). This 10% contributes to reducing the economic viability of the project when producing gas as they will pay the interest deducted by the local government's debt:

> as an investor, we complain. Oh well, we never included this 10% into our calculation when we first took the block. [...] We are already half way through the activities and then we are obliged for this 10% local participation. At first our calculation for our profit was this much, but then we have to deduct our revenue as we need to bear this local government participating interest. (Husin 2018)

Husin (2018) also questioned how the government would contribute to the local area directly with the 10% interest from CBM gas production:

> Ok, the Local Government will get the 10%, and then, how the government will develop the local society? What do they get from the 10% participating interest? Does the Local Government actually hire the local people to work in their office? Right? Do all these local people are hired by the local government, and paid by this 10% interest? No, right? So what is it the Local Government actually contribute to the local area? (Husin 2018)

Notably, the local government is required to pay back the 10% when the investor started producing, with a production share guarantee for the local government (Article 12).

The Involvement 5: Regulated Gas Price

The gas price in Indonesia is regulated under President Regulation No. 40/2016 on Gas Price Determination. The definition of gas price under this regulation is a gas price determined by the minister which has become the basis for the production sharing calculation and the basis for the gas sales price as a result of the implementation of the PSC.

According to regulation PR No. 40/2016 Article 2, there are four main considerations of how the gas price is determined:

- Project economics of a gas field;
- Domestic and international gas price;
- Domestic consumer affordability; and
- Added value of domestic gas used.

For domestic supply contracts, the gas price is negotiated based on a field-by-field basis among SKKMigas, buyers, and individual producers, based on the project economics of a gas field (PR No. 40, 2016). Government distortion on the price is meant to protect domestic consumers from price escalation. Further, this does not imply that the regulated price is set below the economic value; price negotiation between the gas sellers and buyers still happens via a market mechanism. The government's role, then, is to approve the price agreed between them.

The gas price is one of the economic parameters which strongly determines the project economy of a field. Therefore, the government should be concerned with the regulated price, not only to support the industry but also for economic growth in general.

An idea of gas price subsidy mentioned by Gunawan (2018) was that the gas price subsidy could help the economic viability of a project and support the investors. However, this idea was considered unrealistic as the government would still look for revenue.

The Involvement 6: Affordability

One of the CBM investors (interviewees) gave his view on what the government's main priority was at that moment, namely affordability at all costs:

> the main priority of the Government at the moment is affordability at all cost. At all cost. In other words, is how the lowest society can buy electricity at the cheapest price. It's at all cost. (Hamzah 2018)

Affordability relates not only to affordability from the perspective of the customers, but also in the process. The costs to produce gas from CBM must be cheaper so that it can give a low and affordable price for customers. Affordability, then, should be in included in the fiscal regime scenario:

> Affordability in terms of consumer. Yes, that it must be cheaper, to produce the CBM must be cheaper so the customer will get it cheap too. It can be done! In the context of fiscal regime. And this fiscal can work if the nation does not take this as revenue but for development. That is the key.

> Affordability is actually to get it cheaper – this can be done, but the term cannot be 45/55 or not this, but maybe gross split etc. That's the context. So the government shall give it up to the investors, then this industry will work. The industry can work, cos they will get profits from there, and the locals will get it with only $3. Yes it can. [...] So you can create such fiscal regime scenario, the affordability. (Bachtiar 2018)

The Involvement 7: Internal Management Issues

SKKMigas is the only institution that has direct contact with investors in terms of their activities. According to Soeryowibowo (2018), another issue in CBM was the management issue, related to human resources expertise. SKKMigas does not have CBM expertise, and this is the reason why they have treated CBM as conventional gas. This issue is particularly related to human resources skills and expertise, as mentioned by Setyatmoko (2018b). Indeed, Indonesia has no expertise in the CBM, which is not ideal for the current situation.

Readiness

In this regard, the GoI was considered unready to develop a domestic CBM industry. One of the main directors of the IPA, Wajong (2018b), stated that this had been shown by the way the government responded to CBM investors' criticisms and inputs. The government tended to be in a rush when implementing the new fiscal regime and new regulations, adding uncertainties to the CBM investors.

The current situation is that the CBM industry in Indonesia does not work well with the Net PSC, and that Indonesia is not ready to implement the Gross Split PSC for the CBM industry in Indonesia (Wajong 2018b). She mentioned that this new fiscal regime had been implemented without regulation in place to support it. In other words, there is a level of incomprehensibility between the regulation and the new fiscal regime. Even worse, CBM investors who were willing to move from Net PSC to Gross Split PSC were not guaranteed by the regulation. The fact that Indonesia does not have a specific regulation on CBM shows that the government is not yet ready to develop CBM, irrespective of the Net PSC or the Gross Split PSC regime.

4.3.2 Fiscal-Related Issues

Regulation, Procedures, and Fiscal Policy

This section discusses the issues related to the fiscal and regulations attached to the CBM activities. Commerciality is also discussed in this section, due to the failure of fiscal policy and procedures pertaining to CBM in the country.

The GoI conducted a feasibility study on CBM technical and economic aspects from 1998 to 2004 (see Fig. 4.3), until the first pilot project was put in place. There has been a specific regulation assisting the new born CBM industry since 1998, some of which have since been revoked or amended. The GoI classified the CBM industry as part of the upstream oil and gas industry; therefore, the CBM industry must also follow the general rules and regulation applied to the upstream oil and gas industry.

Gunawan (2018), the former head of the Unconventional Hydrocarbon Department at SKKMigas, stated that the Net PSC was quite attractive because of the idea of risk-sharing between the government and the contractor:

> I would say this PSC used to be reasonable for CBM industry. Why not? The split for CBM investors was made higher than in the conventional oil and gas. (Gunawan 2018)

At that time, many investors were attracted to the CBM industry. The first CBM contract was awarded to Consortium Medco and Ephindo Energy in 2008, Sekayu Block, in South Sumatra; followed by another six blocks in the same year. These six blocks were located in Central Sumatra and Kalimantan: one is the Indragiri Hulu Block in Central Sumatra (PT Samantaka Mineral Prima/ CBM Asia Hulu Ltd.), three blocks in East Kalimantan (the Bentian Besar Block, the Kutai Block, and the Sangatta I Block), and two other blocks in South Kalimantan (the Barito Banjar I Block and the Barito Banjar II) (Gunawan 2018).

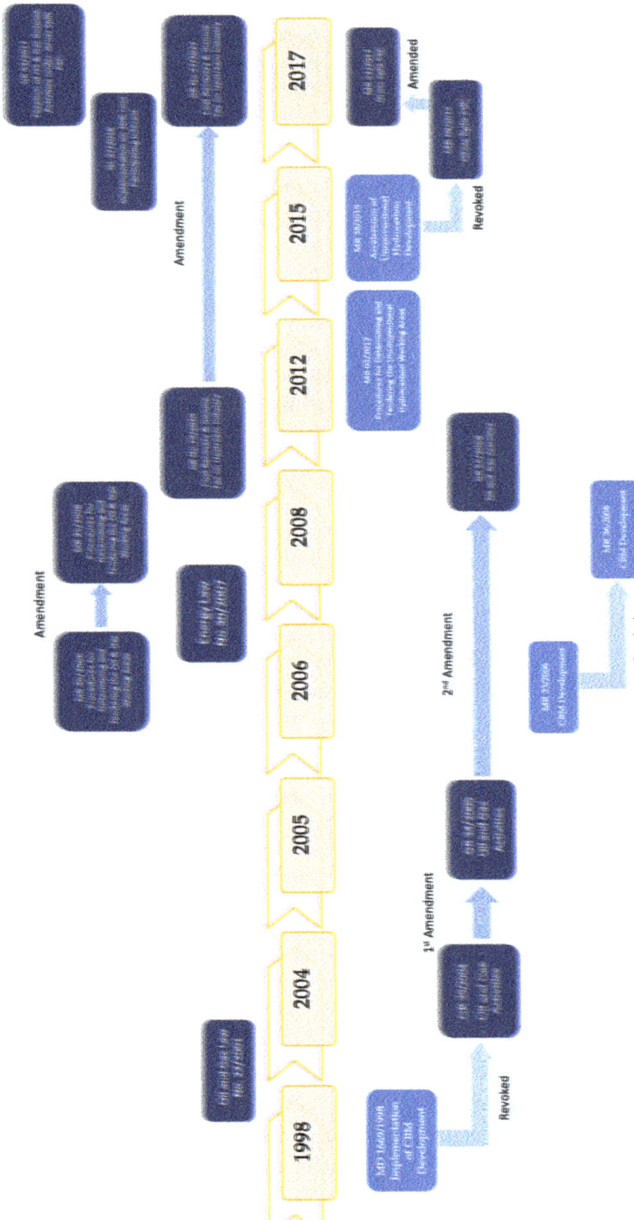

Fig. 4.3 CBM regulation. (Source: Created by the author (2020))

Soeryowibowo (2018), former chairman of the Unconventional Hydrocarbon Committee at IPA, stated a different opinion about the application of the PSC regime to CBM. In his opinion, the PSC was designed only for conventional hydrocarbon. The Net PSC did not accommodate the CBM technical characteristics, which was supposed to be accommodated in the fiscal system, and the fiscal terms did not support the CBM project to be economically viable:

> the technical difference is not accommodated in the current fiscal regime. The current fiscal regime is not supposed to be applied to CBM. So geologically, CBM is very different from conventional hydrocarbon. Unconventional hydrocarbon is different and that is not accommodated in the fiscal calculation. (Soeryowibowo 2018)

Apart from the characteristic differences, one big issue according to him concerned the contract period stated in the PSC. The Net PSC was designed for a 30-year contract, but CBM projects require more than 30 years to reach project economic viability. It is evident that by offering a 30-year contract, the PSC is not attractive for CBM investors, meaning that the government must amend the law to amend the contract period. This will be another significant issue in the CBM industry if the government happens to amend the oil and gas law.

Another significant issue is related to the level of bureaucracy in the government in regards to the establishment of regulations and procedures. CBM technical standards have been equalised with the conventional hydrocarbon technical standard, yet they have different technical requirements; despite this, the government continues to regulate the CBM under conventional hydrocarbon procedures (Golubchenko 2018):

> I think there is the technical standard. Yeah, that has been preventing mmm CBM operators to use cheaper, CBM specific methods or equipment. Yeah, like well barrier, well head requirement. Understand that shallower, lower pressure CBM well could actually using cheaper rate, using lesser well barriers, cheaper well head. But the standard, the technical, the measure of technical standard, has not been adjusted to accommodate that.

Looking at this from different perspectives, Wajong stated that the government should amend the regulation to support the industry to provide the investors flexibility to do their activities. Rahmadhi (2018) also mentioned that flexibility in their activities was better than the ability to recover their sunk costs. As such, Gross Split PSC was considered to be a better fiscal system for CBM as it provided the required flexibility for investors to perform their CBM activities:

> in my opinion, regulation that gives them flexibility. Give them flexibility. So like this, CBM is a new thing. Even if they can switch from net PSC to the new Gross Split PSC, as we all know that Gross Split PSC itself is more flexible and to ease the investors. (Wajong 2018b)

The idea here was to push the costs down, and, according to Utama and Golubchenko (2018), not having cost recovery in the system would possibly make the project more cost-efficient as investors would have the freedom to select the tools and equipment in regards to CBM requirement. A CBM project can be economically viable if the cost can be reduced. It is important to reduce the costs, and having massive development will help to push the cost down (Gunawan 2018). He also stated that the costs could be reduced by increasing the work volumes and using a joint contract. In theory, this is possible, but in reality it would be hard to implement. It requires trust between the suppliers to implement a joint contract, and this trust is usually significantly related to liabilities, for example, in the event of delays for equipment or rigs.

In contrast, Soeryowibowo revealed that both Net PSC and Gross Split PSC did not fit to CBM industry at all. He thought that the main problem was more due to the fiscal system. He stated that there was a need to redesign the fiscal regime applied to CBM (Soeryowibowo 2018), and that this new fiscal regime should be adjusted to the CBM technical requirements (Mair 2018):

> briefly, it does not fit for CBM in Indonesia, and in my opinion both of them do not fit for CBM industry in Indonesia. [...] But, for CBM with average production rate we can say approximately 0.5mmscfd. 0.5mmscfd is the number in which most of the time used for a case reference in CBM

> production. Meaning, to reach 100mmscfd, one third from conventional gas production, how many wells will you need? 200 wells. Meanwhile, to have 200 production well, it will certainly have a high capex as well. Correct? And, our PSC does not distinct them. So in my opinion, this PSC does not match. I have given an input on this to the Ditjen Oil and gas. This PSC does not fit for… These PSCs have been designed for Conventional gas. At the end, it is one fit for all. It cannot be like that. Because they both have very different characters. So in my opinion both of them do not fit. So that's what i think. Both of them do not fit. The PSC cannot be applied to CBM, not even Gross Split PSC fits it. (Soeryowibowo 2018)

Makmun and Sitepu (2012) and Harrington (2016) underlined that most regulations applied to the industry had limited the CBM players to perform CBM exploration in Indonesia. However, Wajong emphasised that it was not about having the regulation enacted or the minister's discretion; rather, it was more about the change of government paradigm on energy, of viewing energy as the engine of growth instead of simply a source of revenue. This paradigm will be discussed further in the next section.

Wajong (2018b) perceived two main problems regarding the CBM industry in Indonesia. First, the implementation of the regulation, considered as impeding CBM activities. As previously discussed, CBM has very distinct characteristics compared to conventional oil and gas, both technically and geologically:

> CBM requires different procedures for its good and service procurement. (Soeryowibowo 2018)

The current regulations are appropriate for conventional oil and gas. Unfortunately, it does not accommodate the differences in unconventional hydrocarbons, such as CBM. Given the circumstances, the CBM players must comply with the conventional oil and gas standard and regulations for tools, equipment and services. Therefore, this has resulted in cost inefficiencies for the CBM players. Wajong further emphasised that same mechanism could not be implemented for two different hydrocarbon 'families' that evidence completely different characteristics. The cost of the initial CBM project was very high due to this.

Even though the Net PSC was considered attractive for its risk-sharing concept (Gunawan 2018), the implementation of Net PSC for CBM is not the best for the project to grow. The technical standard and procedures are very complex, but, importantly, they are not required for CBM activities to run. The Net PSC has procedures, rules and regulation attached to its implementation, and it is these that have become one of the significant issues facing the CBM industry. Instead of supporting the industry to grow, in fact they have hampered the growth of the industry:

> CBM is classified as in oil and gas category, therefore, the standard applied to the CBM industry is the standard which is also applied to the conventional oil and gas, in which, there are many procedures, both technically or other formalities procedures in the fiscal regime for conventional oil and gas. [...] since the very early stage, CBM was classified as oil and gas industry, therefore the standard applied to CBM is similar to the qualification for conventional oil and gas. Whereas, in fact, the CBM could have been simpler and less complex. [...] Implementing the fiscal regime for CBM with the conventional oil and gas standard creates burden for the CBM investors yet at that time, there was inadequate technical support for the CBM industry. At the end of the day, it leaded into the long process due to the permit for the equipment, and secondly the cost was getting higher. (Gunawan 2018)

> Not only fiscal, just regulation and procurement and all kind of stuff which like brings in affect initial projects which are very cost intensive. (Golubchenko 2018)

> Mainly in the Net PSC, it regulates the procurement for goods and services, as well as for land acquisition, and also for procuring the human resources and everything in the WPNB must be approved by the SKKMigas. This is not attractive at all in a way. (Rahmadhi 2018)

As an active CBM investor in Indonesia, Setyatmoko provided his thoughts on how the government has dealt with CBM regulations, namely that they are yet to be improved:

> There has been no regulation to protect us for, not to protect us, but to guarantee that we can apply the proper CBM technical ways. That's one thing that I think Indonesia needs to improve. So yes, the knowledge and understanding are there, but the application is not yet touching the technical sector. (Setyatmoko 2018b)

Under the Net PSC arrangement, CBM investors must also deal with the approval process before conducting their CBM activities at both the exploration and the development stages (Setyatmoko 2018b). This is strongly related to the cost recovery mechanism in place when the Net PSC is applied. Without having these approvals prior to CBM activities, the sunk cost will not be recovered by the government. Moreover, CBM investors are required to obtain both technical approval and budget approval. Technical approval is related to the technical matters for CBM, whilst budget approval is related to the budget for each proposed activity. Plan of Development/Field Development Plant (POD/FDP) approval must be obtained before they can commence production. This POD approval is required in each well that is going to be developed. As mentioned previously, the Net PSC fiscal regime requires a POD for each well, and it is this that has become one of the main challenges facing the CBM industry due to the low production rate of each CBM well. This low production rate also impacts on the number of wells to be drilled to attain commercial value. In the case of CBM, this cannot be applied as it is, since the project requires a hundred wells to be commercialised (Soeryowibowo 2018; Mair 2018). The procedures are also very complex and time consuming:

> Mmm… A hundred wells, if you are doing it by contracts and you have got one two three wells per contract is thirty contract or something. (Mair 2018)

> Hmm Net PSC does not fit the CBM project. It does not fit. It does not fit really. If it is fixed like conventional, it does not fit. It does not fit for CBM. It must be modified. Must be modified. Must be modified. (Lubiantara 2018)

In order to keep up with production rates, constant well drilling activities will be necessary during the exploration and production stage. In other words, there will be constant CAPEX and OPEX during the life cycle of CBM.

The second problem according to Wajong was particularly related to the Gross Split PSC. The Gross Split PSC is regulated under Ministerial Regulation No. 8/2017, replacing MR No. 38/2015. This is one of the challenges related to the regulation and fiscal regime implemented for CBM. The government has not finalised the regulation and procedures for those who want to change their contract from the Net PSC to Gross Split PSC, with no clear instruction or procedures from the current CBM players to move to the Gross Split PSC. The government is yet to be focused on this, particularly for MR No. 8/2017 being replaced by MR 52/2017.

They revoked MR 38/2015, the first regulation established specifically for CBM, and overwrote it with MR No. 8/2017. This is one of the challenges related to the regulation and fiscal regime implemented for CBM. Worse, no one was actually appointed to deal with CBM.

Another significant challenge related to regulation is that of overlapping regulations. For example, the forestry regulation, the law on land and agrarian, environmental law and regulation which overlap with local government regulations and procedures (Fauzi and Anna 2013; Dutu 2016; Setiawan et al. 2016). These are directly related to the initiation of any oil and gas project in Indonesia. This has become even worse when the CBM resources are underneath the coal mine blocks which belong to other investors, or the CBM block is located in a palm plantation owned by palm companies (Setyatmoko 2018b). This creates a more complex situation for CBM investors since they also must deal with other companies and comply with any relevant regulations to get the right to extract the CBM gas. CBM investors have to comply with all these regulations to obtain permits to get the project started.

Commerciality

Selling gas is more difficult than selling oil. To be able to sell gas, the producer is required to initially find the buyer and build the infrastructure, which costs four times more on average than what is comparable for oil (Stickley 2014). According to Stickley (2014), four elements are required for the gas industry to be successful: (1) it requires a large amount of gas to be commercially viable; (2) it requires infrastructure such as pipeline; (3) the gas market has a price constraint which is the case in Indonesia as government regulates the gas price; and (4) how the government can provide the incentives for the activities. According to Golubchenko (2018), a Petroleum Fiscal analyst from WoodMackenzie, it is not easy to find a buyer:

> even for operators with bigger amount of gas, they still find that gas allocation, or finding buyers is very hard. (Golubchenko 2018)

Project commerciality is determined by the revenue and cost of a project. Revenue is attached to the volume and price of the commodity. However, in a gas project, the volume is defined in the gas sales agreement, but the price is either using the market price or the regulated price. In Indonesia, the gas price is regulated by the MoEMR, including for CBM gas. The CBM cost was initially very high due to the PSC and its attached procedures and regulations, although, lately, CBM investors have managed to reduce the costs as they are now able to use a proper technology for CBM operations (Husin 2018). One way to boost CBM commerciality is by increasing the gas price, but it will become a challenge for investors as it will make CBM gas less competitive compared to conventional gas:

> The next problem is the CBM commerciality. CBM commerciality is not easy. However we will sell our gas from CBM, but the buyer may not have the same confidence level as they have when they buy the gas from conventional hydrocarbon. Since it has not been proven yet. CBM is a marginal project, and we are not looking for profit, but its margin, and added with the market that is not sure with its potential. This makes the commerciality even more difficult. […] we expect that the Government understand and then can help this industry to compete with the conventional hydrocarbon, that's what we expect. […] Even if we manage to save the industry, it does not mean we can commercialise the project. (Setyatmoko 2018b)

> in the beginning of 2009/2010 we used mining rig to push the cost down, yet the cost was still very high. I don't understand why. It may be caused by the mobilisation cost ...[...] So, even there has been an innovation for CBM technic, which is one of the key thing, but in the reality, it was still not enough to push the production cost down [...] yes that was because the gas price. If only the gas price was still $14, there will be many investors jump into CBM. (Husin 2018)

NuEnergy Ltd. has a concern regarding the gas market and the competitiveness of CBM gas in the market. At this moment, CBM gas is not very competitive compared to conventional gas, as it is challenging to create a competitive gas price for CBM gas. According to Setyatmoko, since the gas price is regulated, and unless the government is aware and takes action on this, the market will never be competitive for CBM gas. Another challenge is that the CBM gas market is not yet stable, unlike conventional gas. People are more aware of conventional gas than CBM gas (Setyatmoko 2018a). Regardless of all the challenges that they face, NuEnergy Ltd. is the only CBM investor that is moving forward. It is believed that when they prove that their CBM is commerciality viable, there will be an opportunity to revive the industry in Indonesia.

CBM gas tends to be sold and used domestically, mainly due to issues of commerciality, and the second one is due to regulation. According to Golubchenko (2018), there is no problem with selling gas domestically because it is more feasible than selling gas non-domestically because of the limitations due to regulations and politics (Golubchenko 2018). According to Mair (2018):

> But if you are gonna go for domestic gas supply, then you go for the domestic gas price, you can price whatever it works. Whatever it works to reach for that village or, for that power plant or factory then whatever. Then it would work. But there is gonna be flexibility there that you can just make it work for all parties. And that's bit tricky. Because then you have power politics and crisis.
>
> The price will have to compensate for the fiscal system. So you can either say reduce government take or you can increase price. (hahaha) it is just leading the problem to downstream. (Mair 2018)

The investors and government may set the gas price at any price that works for them to reach whatever goals they seek to address, but flexibility is still required to make things work for all parties (Mair 2018).

4.3.3 Local Community–Related Issues

Corporate Social Responsibility and Unpredictable Social Cost

Oil and gas production and exploration activities always directly affect the environment, with many of them negative impacts. These include climate change, loss of habitats and ecosystems and biodiversity loss. Hence, this industry is always tied with the local issues since its existence brings direct impacts to indigenous people, local communities and the environment. There is always a need for the local people to have compensation over the impact of the activities. This can be financial compensations and/or improvement in public facilities, such as healthcare facilities, roads and other infrastructure to improve their quality of life (Freilich and Popowitz 2012). These activities are often called corporate social responsibility (CSR) nowadays (Frynas 2009).

Unlike in Australia and the US, CBM areas in Indonesia are mostly in densely populated areas; therefore, the CBM players should take social costs into consideration. Some CBM blocks are overlapping with the palm plantation or coal mining blocks. It is crucial for the CBM players to be aware of the hidden cost caused by the social and environment where their blocks are located (Soetrisno 2018). The author believes that the CBM players should have contingency funds or allowances for social costs since these costs are unpredictable.

Soetrisno (2018) emphasised that social costs can impact directly on the investment decision of prospective investors. CBM investors can calculate the estimated reserves and evaluate the project economy by using the existing data, but there is a social cost that is unpredictable and can bring the economics of a project down and eventually create an uneconomical project. This shows that it is crucial for CBM investors to have a CSR fund and financial allowance to cover these unpredictable social costs as well as a good corporate behaviour towards society and environment.

The European Commission defines CSR as a concept whereby companies integrate social and environmental concerns into their business activities, as well as their interaction with their stakeholders. CSR is to be voluntary (European Commissions 2001). Frynas (2009) found that it was important for local government to be involved in CSRs programmes. According to Freilich and Popowitz (2012), local government has a role in guaranteeing the health, safety and environment in their area. CSR can also bring long-term benefits for investors as it could also potentially enhance their public image (Streimikiene et al. 2009).

SAGE (2012) stated that corporate social performance (CSP) is an extension concept of CSR that emphasises on the achieved results rather than just a responsibility towards societies. A study by Orlitzky et al. (2003) proved that CSP has a positive relationship with the corporate financial performance of the institution. A second-order meta-analysis on this relationship by Busch and Friede (2018) claimed a highly significant, positive, and bidirectional relationship between the CSP and the corporate financial performance.

It is clear that this expenditure component is very important for the company's sustainability; therefore, this research will focus on this cost as one of the main terms in the fiscal model being constructed.

The Government and Investors Views on Local Issues

CBM operations always involve fracturing (Rodvelt 2014), and it is believed that this activity damages the environment (Meng 2017; Mayer 2016).[4] Therefore, it is important for CBM players to implement CSR policy and to have a financial allowance to cover the negative impacts of their activities as mentioned previously.

[4] Many literatures being stipulated in this chapter are related to the fracking and its impact on the environment. Impacts that are related with nuisance like noise, traffic, and dust (Jacquet 2012; Evensen et al. 2014; Willow 2015), impacts on water pollution (Theodori 2009; Ladd 2013; Jacquet and Stedman 2013; Brasier et al. 2013), and impacts on the public health (Kriesky et al. 2013; Jalbert et al. 2014).

A CBM project in Indonesia is very much related to the indigenous people, and it can be assured that there will be constant interaction amongst them.[5] Hence, it is important for them to invest in the development of the local communities to support their CBM operations with less friction (Frynas 2009). It is also to be understood that this interaction requires good management from investor's side to ensure that there is positive interaction in the process. As suggested by Sage (2012) and Patari et al. (2014) on the relationship of social costs and financial performance in the energy industry, having CSR activities can potentially result in cost savings as it will reduce the unnecessary social costs by having a better understanding with the local communities and therefore reduce the risks of conflicts with them.

Local communities usually already have high expectations on any industry activities that are being carried out in their area. As an investor, they have anticipated these expectations; for example, Soeryowibowo (2018) mentioned all the possible expectations by the local people. The local people may be expecting employment rates to increase, that the economy will grow, better electricity supply, and a better quality of life in general.

The CBM investor, NuEnergy Ltd., has overcome the challenges in relations to local people. They managed to communicate well with the locals and prepared them to accept their project as part of their project to benefit them. Prior to production, communication between the investor and local communities is the earliest and easiest way of getting support from the locals:

> It is not disturbance but it is the challenges. Overall, we could communicate very well, we could have a good relations with them. […] and then we prepared their mentality to also take this project as part of theirs. Cos this bring great benefits for them as well. (Setyatmoko 2018b)

In fact, local involvement is very important to help the industry move forward (Wajong 2018b). The investors should consider and think about dealing with the local communities as these people face direct impacts

[5] Author's survey in CBM development areas in 2017–2018.

from their activities. However, it is important that local people are the ones who define what they need from the industry, what they expect from the industry and define what the industry can do for them as well. For example, the local people could encourage the government that their area should get supply from the investor locally since the activities of the industry will also accelerate the economic growth in their area. It is more about a collaboration between the local area and the industry:

> The thing is that the CBM wells are located in such areas, meaning, they need the locals. So, I think, the idea, the involvement of the local is the most accurate idea as the locals can help the industry directly. Secondly the local area, what Unggul (Setyatmoko) said about commerciality, please prioritise that. So in this case, the local will help the industry by telling the government, use this one, use the CBM gas, since it is also for them. That way, we don't need to push the government. If the CBM wells are located close to their area, the best thing to do is to utilise the gas for that local area. So now they say, I want to use it for this, this, and this. PLN, you use theirs. For my area, take from here. That means, the Government should forget the fact what they are getting, but it is more the collaboration between the investors and the local area. This is the only way to make it work, otherwise, it is impossible to work it out, and then will confuse the commerciality. […] so apart from giving freedom for the industry to deal with their activities, they are also given freedom to collaborate with the locals in which their CBM is developed. Imagine, if this works in many places, it will be very beautiful for the locals, what they are getting, yes of course by also involving into the CBM work. […] so it is a direct impact, and the central Government not to expect anything from the industry. (Wajong 2018b)

The ADPM should help local people regarding this matter. The changes of paradigm cannot only happen in the central government but also in the local government. Local people should be the one stating that they want to use the gas from the CBM project for their area. They should be given a view on what is happening and asked how they will react to it (Bachtiar 2018).

Looking at the CBM activities and the areas required, Utama and Golubchenko (2018) expected that there would be significant impacts on local people. However, they thought of positive impacts rather than negative impacts:

> mmm… unfortunately I haven't noticed much, just because the activity level has not been that high, but, you know, CBM requires at least in the initial years, you know, it's a high, aa sorry it's an intensive, quite an intensive development right. I expect quite significant local impact just because you know, you need to keep on drilling well, and lots of wells are required, land opening, making the gathering, path line. We expected to be positive impact. (Utama and Golubchenko 2018)

According to Utama and Golubchenko (2018), CBM projects will support the economic growth in the areas where these projects take place. This also will bring more employment opportunities to the local people. Understanding the local expectation and what impacts can be brought from the CBM activities is considered very important:

> You see as previously said that the Government is thinking about how much money will be earned, they are getting from the contractor? But if they look more, let's say big picture, in a big picture they don't realise that any promotion to the any industry, and we are talking about CBM now, will boost at least the provincial economic where these projects are taking place, and also will increase the number of place working for this project etc. And all this services will be required you know, that you have these projects. It will basically grow the economy in this provinces, despite that for the particularly upstream sectors, they want just much money for example right? I think that's why it's important mmm … this local is considered is very important. (Utama and Golubchenko 2018)

Rahmadhi (2018) stated that one of their coal mining project areas, Kota Baru, brought positive impact to the local community, bringing economic activities to the community. He was convinced that by looking at their past experience on the coal mining project, the CBM project would bring more positive impacts to this area. This was also agreed by Lubiantara (2018), a local consultant and a government member, that this will bring more economic activities to the local areas. He also stated about the multiplier effect on having industries activities in the local areas.

According to Husin (2018), it is possible to hire local people, but then they have to be aware that working in the industry requires skills and knowledge about the industry:

> But we cannot. I meant, even the CBM operation is smaller than the conventional oil and gas, we just cannot. There are some operational that it is 90% probability is more to technical matters. So that's why, we cannot hire local people for technical operations, yet we still have to consider the safety and we need to train them. So it cannot be done. (Husin 2018)

Therefore, there are not many such employment opportunities for local people. He also stated that hiring local people for technical matters would cost extra money, and yet it takes years to get them to a certain level to be able working in the technical operations. According to Husin (2018), the company could offer local people to deal with the Drilling Site Preparations (DSP), and maybe hire them as security/field guards. This is one of the ways to secure their CBM operations. By hiring local people, it is expected that the communities will ease the process and not disturb the activities being done in their areas:

> And we cannot also spend extra money to train them, cos it will need years, to train until certain level, where they can contribute in technical matters. In the operation, it cannot be done. So what we usually hire in CBM for local people in our CBM area is, that we hire them for dealing with the land preparation. So the DSP, Drilling Site Preparation they are involved. So they, cos it is simple. Yes, we need heavy equipment but we still need to prepare the site. The land, the trees logging, they can participate. But apart from that, it is difficult. Maybe one thing, as a security / field guard. It is doable. Security is possible. And we always have strategy, how we can operate with the local people working for us. Because it will secure our activities. If there is someone they know in our company, they will not dare to disturb our operations. […] so yes, that's our strategy. During the operation, even after we are done with the site preparation, after the operation, we still hire the local people. Cos we need, workers, to lift stuff, for security, that we always involve them. How many we can hire from there? it won't be that many, but that's the way. So they will get the direct impact. Even we

> haven't start our production, but they already received income from us. So there has been impact even in the exploration stage. (Husin 2018)

Similar to Husin, Purba (2018) thinks that it will be difficult for investors to accommodate the local's expectation. This is most likely due to the time required for the CBM project to get into the commercialisation stage. The CBM project takes much longer for it to get to the commercialisation stage compared to other coal mining project:

> The CBM Projects by nature is always neighbouring to Coal Project. Coal contractual does not govern by PSC. It is under general mining. Relatively easier to deal with surrounding communities in term of supporting community development etc.
>
> While such kind of luxury, you don't find in CBM. You may need more than 10 years, still no clear whether you can go to commercialization stage. So it will be difficult for investor to deal with and accommodate the surrounding society's expectation. (Purba 2018)

As a government party, Purba (2018) emphasised how the GoI has dealt with local communities. The GoI has a standardised approach: together with investors they develop a campaign about the project that is being conducted in the local area and how it will develop the local community (community development). This is conducted by cooperating with local government and local community leaders.

4.3.4 Investor-Related Issues

Financial Capability

There is no CBM project being developed today, and as long as there is no proven project, it will not be possible to have new investors coming. Most current CBM investors now find themselves in a difficult financial position as they have typically spent a huge amount of money when they first started:

> to be honest, our investors have stopped to fund our project. We have no more money to continue our project. (Husin 2018)

The investors face financial challenges after spending all their money at the initial stage:

> to be frank, from my company, I don't have the shareholders support anymore to give me more money. That's my problem. That's my problem. Therefore, I have many meetings with CBM investors, and I have told them that we will fight for this industry. Just to make it alive and move, like that. But for me, honestly, my shareholders – you are mad, how many years has it been. G&A how much million dollar? (Hamzah 2018)

The early stage requires a lot of trial-and-error activities, and these trials and errors required lots of money. Therefore, it is crucial for the investors to be cost-efficient:

> It is clear that financial is a big deal in the CBM industry as at this stage, CBM project is still in a very early stage, a stage where investors do trials and errors. They constantly spend their money for this. That's why, it is very important for the investors to be cost efficient. There is no shortcut, everyone must be patient. And cost efficiency is absolute. (Wajong 2018b)

The current challenges faced by CBM investors are the forced termination and its penalty. These challenges are mainly due to their lack of financial capability:

> Mostly due to the lack of financial capability, kept postponing, until they ran out of time, panicking, and at last they asked for penalty relief or replacement. It is not possible. So at the end they asked for addendum. This is terrible, so once the termination letter is out, they must pay the penalty. (Gunawan 2018)

So far, this part has discussed the government-related issues, fiscal issues, local community–related issues, and investor-related issues. The next part will discuss the technical issues related to CBM development.

4.3.5 Technical Issues

Expectation Vs Reality

However, a significant gap was found between expected and determined results regarding the CBM potential. The gas content in the coal was not as high as expected (Rahmadhi 2018):

> And then what happened was, that we have done the exploration activities, and the result technically was not close enough with the result from the joint study we previously did. (Setyatmoko 2018b)

> The result is not as good as we expected before. (Setyatmoko 2018b)

> So what happened was that from the joint study result, it was found that they have potential on CBM, but then when we explored them, they are not potential at all, for example at Kuala Kapuas Block. (Rahmadhi 2018)

> Technically, possibly, in details, at this moment, we have not managed yet to let the gas flows continuously. [...] and we have the same issues as mentioned by Setyatmoko. (Rahmadhi 2018)

> the continuity production from the coal, the gas flow is not as we expected. (Rahmadhi 2018)

> The gas is not big enough and too many technical problems. Basic technic only (Pujobroto, 2018). [...] meaning, this is smaller from the small amount. So you can imagine, the gas being taken will be very small and it will take very long time to get it out. So that's because of its low gas content. (Pujobroto 2018)

As previously mentioned, the investors rushed to get the CBM blocks offered by the government; yet there was inadequate research on the CBM properties. Wajong (2018b) suspected that a false belief on CBM characteristic triggered the initiation of CBM development in Indonesia. This is highlighted by zero progress on CBM projects carried out in

Indonesia, except the one CBM investor currently awaiting POD approval:

> that in fact the CBM in Indonesia is still at the very early stage. Therefore, we still cannot know the potential of the CBM in Indonesia. There is no clear result. Maybe that was because many people thought that the CBM in Indonesia has similarities with the one in Australia or their places. In fact, it is not the same. So the CBM industry in Indonesia is faltered from the technical point of view. (Wajong 2018b)

Dewatering Management

Dewatering management has been a concern in the CBM industry. Apart from the wet coal, the problem that may exist is where the projects can get the water to be injected for CBM fracking purposes. The only possible water source that can be used is the formation of water, as it is not possible to use groundwater (Soetrisno 2018). Soetrisno (2018) stated that this situation had created another social problem. The government and the social community have no knowledge on this, so they sometimes charged the investors a social cost towards the use of groundwater or for contaminating the groundwater.

However, water injection also leads to another technical problem since Indonesian coal is very wet. This means that CBM is technically different compared to conventional hydrocarbon developments, and this condition has been one of the main challenges to producing gas from CBM:

> But the big problem is a lot of the early explorers require them waters their wells with water out very quickly, it is just the whole lots of water in the system. A very wet coal. And mmm so they would struggle [...]. So the quality of coal in that regard which is a challenge. (Mair 2018)

> so water can be too little but it's I think in Indonesia is a lot. And I think the comments are much common challenge is the water management is hard. Mmm... It's too much water in the coal. Too watery in the coal. Takes a long time. And then there is not much gas... (giggling). (Mair 2018)

Lubiantara (2018) emphasised that the dewatering process is the one process that occurs in CBM projects but not in conventional hydrocarbon projects. It is uncertain when the dewatering process in Indonesian's CBM project will end since the coal contains so much water; it takes a very long time to finish the dewatering process, and it is tough to predict how much time it will take overall.

Coal Characteristics

According to Soetrisno (2018), CBM in Indonesia remains unproven, and Indonesian coal beds are not thick enough for CBM development:

> Technically, the CBM in Indonesia is not proven yet, since our coal's layer is thin compare to other countries. (Soetrisno 2018)

This was also confirmed by Pujobroto (2018), a former CBM exploration manager of the biggest owner of CBM blocks in Indonesia. He mentioned that there were many technical issues that contributed to the failures of the CBM industry in Indonesia:

> The problem is, so we are looking at gas content and permeability. The deeper it is, the higher is the gas content, the bigger this gas content. It will be gone and down again. In reverse, the permeability, it is the contrary of the gas content. The closer to the surface, the permeability is bigger, but the deeper it is the lower the permeability. So this is what makes does not work sometimes. For example, in Barito Basin. This area shows that the gas content can reach to 200scf, but how deep it is? 1300 meters, but the permeability is 0 in this depth, the gas cannot be out anymore since the permeability in the coal can't be forced. We cannot force it. So when the permeability is 0, and the gas content is high, it still cannot be taken. This means that we have to find the optimal gap to get the gas out. [...] this is what makes it does not work, it is its geology. Our geology, low gas content and limited permeability. (Pujobroto 2018)

Here three main issues are described:

(a) The coal is distributed widely, and the structures are different even if located in the same area.
(b) The rank of the coal is too low to have a good gas content.
(c) It has low permeability. This permeability is related to the gas content itself. The depth has a strong connection with how high the gas content will be and the level of permeability. One rule of thumb is that the deeper the coal, the higher is the gas content. In contrast, the deeper is the coal, the lower is the level of permeability.

Gunawan (2018), the former head of unconventional hydrocarbon division at the SKKMigas, said that CBM was less risky compared to conventional hydrocarbons, as the CBM resources tended to be spread all over the place. Thus, it was not as technically structural as in conventional hydrocarbons. According to him, investors should be encouraged more, and be braver, to invest in the CBM industry:

> Actually this CBM is technically, from the risk perspective, it has lower risk compared to conventional hydrocarbon. In conventional hydrocarbon, when we play only in one prospect or lead, in which we drill, we will find out whether it dryhole or not. Meanwhile in unconventional hydrocarbon, it is non-structural. Once we successfully prove that the property underneath is considered commercial, then it will be easier because it has wide range of distribution. [...] very distinct to conventional hydrocarbon that is very spotted. [...] Once they prove that the deliverability of the property is good, commercial, they will continue to run the project. [...] Technically, the risk is very much lower compare to conventional hydrocarbon. Therefore, the chance to be successful and to get the cost recovered, so this means that the people should be braver to invest in CBM than in conventional hydrocarbon. That was why, the CBM investor preferred to have PSC. (Gunawan 2018)

In addition, Gunawan (2018) stated that coal permeability in the South Sumatra Basin was higher than that found in the Kalimantan, Kutai and Barito basins. Another concern about the CBM gas is that of the fast decline rates of CBM gas in each well. Therefore, hundreds of wells are required just to keep up the production profile rate. Surprisingly, he is convinced that Indonesia has an adequate coal thickness and gas content in the CBM basins, even though he mentioned that the permeability might not be good enough for CBM gas to be extracted.

Technology and Infrastructure

The problems of transporting the CBM gas and the connectivity between the islands is another challenge, and one that has become increasingly complex since there exists inadequate infrastructure in place to ease transport concerns. However, since CBM is planned to be locally used, this will not create a problem regarding distribution of the production. This is much more related to the material, tools and equipment mobilisation, and it can still increase the risks and costs of mobilisation (Lubiantara 2018):

> the challenge in Indonesia is the infrastructure. There is no infrastructure availability which distinct this with the US. the infrastructure availability. The second one is in Indonesia, why is it difficult? It is because the connectivity, we are different with the mainland. [...] Indonesia is an archipelago country.

The equipment standard in the Net PSC applied to conventional oil and gas is very costly, yet the CBM project is required to follow the standard, regardless of the characteristic differences between conventional and unconventional hydrocarbons:

> it's about the equipment. The equipment we are using from the early stage, drilling stage til the pumping, it's all with conventional oil and gas standard, and it is very expensive. (Rahmadhi 2018)

According to Gunawan (2018), the government changed its requirement for procuring CBM tools and equipment. It has now become cheaper, and investors should take advantage of the remedied situation. However, the initial massive costs that they spent during the early stage of CBM development means that they are now facing financial difficulties.

References

Bachriadi, D. and Suryana, E., 2016. Land grabbing and speculation for energy business: a case study of ExxonMobil in East Java, Indonesia. [online] Available at: <http://dx.doi.org/10.1080/02255189.2016.1197825> [Accessed 20 February 2020]

Bachtiar, A., 2018. *A Focus Group Discussion on CBM Industry in Indonesia.*

Brasier, K., McLaughlin, D., Rhubart, D., Stedman, R., Filteau, M. and Jacquet, J., 2013. Research Articles: Risk Perceptions of Natural Gas Development in the Marcellus Shale. *Environmental Practice*, 15(2), pp. 108–122.

Bryner, G., 2002. *Coalbed Methane Development in the Intermountain West: Primer*. Colorado: University of Colorado Law School.

Busch, T. and Friede, G., 2018. The Robustness of the Corporate Social and Financial Performance Relation: A Second-Order Meta-Analysis. *Corporate Social Responsibility and Environmental Management*, 25(4), pp. 583–608.

Cully, M., 2015. *Review of the Socioeconomic Impacts of Coal Seam Gas in Queensland*. Canberra: Australian Government, Department of Industry, Innovation, and Science.

Dutu, R., 2016. Challenges and policies in Indonesia's energy sector. *Energy Policy*, 98, pp. 513–519.

Evensen, D., Jacquet, J., Clarke, C. and Stedman, R., 2014. What's the 'fracking' problem? One word can't say it all. *The Extractive Industries and Society*, 1(2), pp. 130–136.

Fauzi, A. and Anna, Z., 2013. The complexity of the institution of payment for environmental services: A case study of two Indonesian PES schemes. *Ecosystem Services*, 6, pp. 54–63.

Freilich, R. and Popowitz, N., 2012. Oil and Gas Fracking: State and Federal Regulation Does Not Preempt Needed Local Government Regulation. *The Urban Law*, 44(3), pp. 534–575.

Frynas, J., 2009. Corporate social responsibility in the oil and gas sector. *The Journal of World Energy Law & Business*, 2(3), pp. 178–195.

Golubchenko, N., 2018. *Views on CBM Industry in Indonesia (Global Consultancy's Perspective).*

Gunawan, B., 2018. *Views on CBM Industry in Indonesia (Government's Perspective).*

Gylfason, T. and Zoega, G., 2006. Natural Resources and Economic Growth: The Role of Investment. *The World Economy*, 29(3), pp. 1091–1115.

Hamzah, S., 2018. *A Focus Group Discussion on CBM Industry in Indonesia.*

Harrington, J., 2016. CBM Indonesia – Dull Past, Bright Future. In: *2016 Technical Symposium*. Jakarta: Indonesia Petroleum Association.
Hulme, D., 2002. *Coalbed Methane Development in Wyoming's Powder River Basin*. [ebook] Colorado: University of Colorado Law School. Available at: http://scholar.law.colorado.edu/coalbed-methane-development-intermountain-west/4 [Accessed 19 March 2020].
Husin, M., 2018. *Views on CBM Industry in Indonesia (Investor's Perspective)*.
Jacquet, J. and Stedman, R., 2013. Perceived Impacts from Wind Farm and Natural Gas Development in Northern Pennsylvania. *Rural Sociology*, 78(4), pp. 450–472.
Jacquet, J., 2012. Landowner attitudes toward natural gas and wind farm development in northern Pennsylvania. *Energy Policy*, 50, pp. 677–688.
Jalbert, K., Kinchy, A. and Perry, S., 2014. Civil society research and Marcellus Shale natural gas development: results of a survey of volunteer water monitoring organizations. *Journal of Environmental Studies and Sciences*, 4(1), pp. 78–86.
Kriesky, J., Goldstein, B., Zell, K. and Beach, S., 2013. Differing opinions about natural gas drilling in two adjacent counties with different levels of drilling activity. *Energy Policy*, 58, pp. 228–236.
Ladd, A., 2013. *Stakeholder Perceptions of Socioenvironmental Impacts for Unconventional Natural Gas Development and Hydraulic Fracturing in the Haynesville Shale*. Texas: Southern Rural Sociological Association, pp. 56–89.
Legowo, E., 2018. *Views on CBM Industry in Indonesia (Government's Perspective)*.
Letts, L., 2012. *Coal Seam Gas Production – Friend Or Foe of Queensland's Water Resources?*. 1st ed. Australia: Thomson Reuters.
Lubiantara, B., 2017. *Paradigma Baru: Pengelolaan Sektor Hulu Migas Dan Ketahanan Energi – New Paradigm: Managing the Upstream Industry and Energy Security*. Jakarta: Grasindo.
Lubiantara, B., 2018. *Views on CBM Industry in Indonesia (Local Consultancy's & Government's Perspective)*.
MacKinnon, A. and Fox, K., 2006. Demanding Beneficial Use: Opportunities and Obligations for Wyoming Regulators in Coalbed Methane. *Wyoming Law Review*, [online] 6(2), pp. 369–400. Available at: https://scholarship.law.uwyo.edu/cgi/viewcontent.cgi?article=1135&context=wlr [Accessed 19 March 2020].
Mair, D., 2018. *Views on CBM Industry in Indonesia (Global Consultancy's Perspective)*.
Makmun and Sitepu, E., 2012. Makmun. *Effect of fiscal incentives on coal bed methane price: a hypothetical analysis*, 3(1).

Mayer, A., 2016. Risk and benefits in a fracking boom: Evidence from Colorado. *The Extractive Industries and Society*, 3(3), pp. 744–753.

Meng, Q., 2017. The impacts of fracking on the environment: A total environmental study paradigm. *Science of the Total Environment*, 580, pp. 953–957.

Mudd, M., 2012. Montana v. Wyoming: An Opportunity to Right the Course for Coalbed Methane Development and Prior Appropriation. *Golden Gate University Environmental Law Journal*, 5(2), pp. 297–342.

Noor, R., 2015. Analisis Kebijakan Gas Metana Batubara Indonesia Saat Ini: "Analysis on the Current CBM Policy in Indonesia)." *Kompasiana*, [online] Available at: https://www.kompasiana.com/ryanalfian/55009f6ca33311ac0a5100d3/analisis-kebijakan-gas-metana-batubara-indonesia-saat-ini [Accessed 19 February 2020].

Orlitzky, M., Schmidt, F. and Rynes, S., 2003. Corporate Social and Financial Performance: A Meta-Analysis. *Organization Studies*, 24(3), pp. 403–441.

Pätäri, S., Arminen, H., Tuppura, A. and Jantunen, A., 2014. Competitive and responsible? The relationship between corporate social and financial performance in the energy sector. *Renewable and Sustainable Energy Reviews*, 37, pp. 142–154.

Pujobroto, A., 2018. *Views on CBM Industry in Indonesia (Investor's Perspective)*.

Purba, S., 2018. *Views on CBM Industry in Indonesia (Government Perspective, Written Interview)*.

Rahmadhi, H., 2018. *A Focus Group Discussion on CBM Industry in Indonesia*.

Rodvelt, G., 2014. Vertical Well Construction and Hydraulic Fracturing for CBM Completions. *Coal Bed Methane*, pp. 101–135.

SAGE, 2012. Corporate Social Responsibility and Corporate Social Performance. In: *Sage Brief Guide to Business Ethics*. [online] Thousand Oaks: Sage Publisher Inc., pp. 231–248. Available at: https://doi.org/10.4135/9781452243979.n26 [Accessed 16 September 2018].

Setiawan, E., Maryudi, A., Purwanto, R. and Lele, G., 2016. Opposing interests in the legalization of non-procedural forest conversion to oil palm in Central Kalimantan, Indonesia. *Land Use Policy*, 58, pp. 472–481.

Setyatmoko, U., 2018a. *A Focus Group Discussion on CBM Industry in Indonesia*.

Setyatmoko, U., 2018b. *Views on CBM Industry in Indonesia (Investor's Perspective)*.

SKKMigas, 2018. *Coal Bed Methane Blocks in Indonesia 2008–2018*. Jakarta: SKKMigas.

Soeryowibowo, M., 2018. Views on CBM Industry in Indonesia (Investor's and Industry Association's Perspective).

Soetrisno, S., 2018. *Views on CBM Industry in Indonesia (Industry Observer's Perspective).*

Stickley, D., 2014. *A Framework for Negotiating and Managing Gas Industry Contracts*. 2nd ed. Dundee.

Streimikiene, D., Simanaiciene, Z. and Kovaliov, R., 2009. Corporate Social Responsibility for Implementation of Sustainable Energy Development in Baltic States. *Renewable and Sustainable Energy Reviews*, 13, pp. 813–824.

Theodori, G., 2009. *Paradoxical Perceptions of Problems Associated With Unconventional Natural Gas Development*. Texas: Southern Rural Sociological Association, pp. 97–117.

Utama, J. and Golubchenko, N., 2018. *Views on CBM Industry in Indonesia (Global Consultancy's Perspective).*

Wajong, M., 2018a. *A Focus Group Discussion on CBM Industry in Indonesia.*

Wajong, M., 2018b. *Views on CBM Industry in Indonesia (Industry Association's Perspective).*

Willow, A., 2015. *Wells and Well-Being: Neoliberalism and Holistic Sustainability in the Shale Energy Debate*. [online] Taylor & Francis. Available at: http://www.tandfonline.com/doi/full/10.1080/13549839.2015.1017808 [Accessed 9 February 2020].

5

Environmental Impacts and Societal Justice Aspects of CBM in Transitioning to Low-Carbon Economy

Abstract CBM projects are always located onshore, and this chapter discusses how CBM projects have impacted local communities both economically and environmentally. A survey on the local impacts and local community's expectation was conducted to collect the data for this purpose. This chapter will also elaborate further in terms of energy justice issues, which will be explored further in regard to CBM development for example what do they want from the project apart from energy supplies for the community (i.e. distributive, procedural, recognition, and restorative justice)? Therefore, this chapter explores in detail the impacts of project development on local community development and how the impacts and benefits can be distributed fairly to the community.

Keywords Coal bed methane • CBM challenges • Paradigm • Regulation • Fiscal issues • Government • Local community • Fiscal policy • Commerciality • Indonesia

T. B. Sumarno, *Just Energy Transitions and Coal Bed Methane*, Energy, Climate and the Environment, https://doi.org/10.1007/978-3-030-85490-4_5

5.1 Introduction

It is commonly known that all energy projects have an impact on the environment and society in general. CBM projects are no different, and the question for government and investors is just how significant is the impact? One of the main activities in CBM projects is fracturing, which directly impacts the environment, especially land (soil) and groundwater. CBM projects are always located onshore, and this chapter provides an analysis of how CBM projects have impacted local communities both economically and environmentally. A survey of the local impacts and local community's expectation was conducted to collect the data for this purpose. Further, this chapter will elaborate on energy justice issues, with a focus on CBM development, that is, what do they want from the project apart from energy supply for the community (i.e. procedural, distributive, recognition, and restorative)? Therefore, this chapter explores the impacts of project development on local community development and how the impacts and benefits can be distributed fairly to and across the community and, subsequently, society.

5.2 CBM's Environmental Impacts

CBM investors need to be responsible towards the local communities in which the CBM activities are being conducted. They need to compensate for all disturbances as a result of such activities, including the construction of roads, well pads, pipelines, and other facilities (Fisher 2016). CBM development can affect the land, water, wildlife, and local communities in many ways. This development can also bring both positive and negative impacts on the areas.

As a positive impact, CBM development benefits local economic growth (Waeckerlin 2005; Mudd 2012; Mackinnon and Fox 2006; Hulme 2002; Bryner 2002; Letts 2012; Cully 2015). It boosts the economic activities in those areas, and it can also improve the local government income. However, local government needs to be aware of other impacts, especially the negative impacts (Bryner 2002; Skov and Myers

2004; Letts 2012; Mudd 2012; Cully 2015). As a community being impacted directly from the CBM development, they have the rights to get compensated for its negative impacts. It is important for investors to prevent disturbances and reduce their impacts.

CBM development has been regarded to have more significant impacts on the environment and community compared to conventional oil and gas development. Despite the number of wells required per project, Indonesia applies the same fiscal regime as applied to conventional hydrocarbons, which have a different level of impact on the environment and community. These environmental impacts have a significant influence on the life of the local communities, and therefore it is important to take these impacts into account in formulating the fiscal regime.

(i) **Water disposal**

Husen (2008) discussed CBM water challenges. In particular, dealing with the water produced from CBM wells has been a critical challenge in the CBM industry, and managing the water produced in this industry is related to economics and environment, both part of practical considerations of CBM producers.

Design and process challenges are two of the main challenges in CBM water management. The design challenges are related to the water flow rate, which is varied, requiring scalable solutions to fulfil the requirements of the treatment facilities being installed for the project. The process challenges are very much defined by the quality of the produced water and waste water management economy. There is a case where water must be reinjected to the formation. Therefore, it is important to minimise waste to reduce water management costs, which has become the main objective in designing the treatment process (Husen 2008).

Hartiniati (2016) also noted that one of the main environmental challenges of CBM activities is water disposal. The disposed water may contain high total dissolved solids (TDS) such as nitrate, nitrite, chloride, or organic compounds (benzene, toluene, ethylbenzene). This composition may differ depending on the coal. As Husen emphasised, water disposal management is a critical concern for project continuity.

(ii) Groundwater

In some cases, for example areas with limited groundwater resources, the CBM project brings a significant impact to the environment due to its water production activities. As discussed previously, CBM gas is produced by changing the underground pressure by pumping the groundwater to the surface and, at the same time, the gas flows to the production well (Hartiniati 2016).

This activity reduces the amount of the groundwater resources for local people, and the water may be contaminated. However, according to the interview in January 2018, Soetrisno (2018) explained that investors were not allowed to use groundwater for hydraulic fracturing. In this case, then, the groundwater issue may not apply to Indonesian CBM.

(iii) Land surface

CBM development also brings a direct impact on the land surface (Skov and Myers 2004; Mackinnon and Fox 2006; Hulme 2002; Bryner 2002). As previously discussed, a CBM project requires an extensive area to drill multiple numbers of wells. In a very dense population area, this will significantly influence community life. Land permits from local government are crucial to start a CBM project (Government Regulation No. 35/2004). Investors are required to obtain land permits before starting any activities, and, consequently, it has to be done at the earliest stage immediately after the government awards the block.

Plenty of roads will be required to access to development area for equipment and tools as well as for the workers (Setyatmoko 2018).

Developing facilities for CBM activities also bring direct impacts on the environment. Native vegetation, wild animals, or any groundwater sources, as well as nearby farming operations, maybe influenced by the activities (Hulme 2002; Bryner 2002). These activities may also generate negative reports concerning CBM development (Hulme 2002; Bryner 2002). Importantly, the population in CBM development areas will increase due to the number of workers required for each project. Although this will drive the local economy, public infrastructure will need to be built to fulfil the increasing demand, which can be beneficial for the community.

(iv) **Air pollution**

Flaring and venting contribute to climate change as it directly releases emissions into the atmosphere (Emam 2015) and pollute the air which will impact human health and the environment (The Pembina Institutes 2003). Like any other gas project, CBM involves flaring and venting, which contributes to air pollution and climate change. Despite the amount of gas being flared in the CBM project, flaring and venting in the CBM project will take a much longer time compared to conventional gas wells (The Pembina Institute 2003).

CBM development field accessibility is also required for the CBM activities to run. This access is much related to the transport of CBM equipment and workers, which can also produce emissions and contaminate the air (Hulme 2002). The polluted air will contribute to the health problems of local communities as well as animals and plantations (Bryner 2002; Hulme 2002).

The minister of national development planning has been aware of the problem related to gas flaring. According to the minister, the regulation for gas flaring can be categorised into two areas: regulatory and operational. Non-technical matters such as application and procedure, measurement and reporting, as well as monitoring enforcement are covered in the regulatory category. Technical aspects such as the timing and location of flaring and venting, burn practices and technology, smoke and odour, and heat and noise generation are covered under the operational category. This GHG emissions standard has been set under the Minister of Environment Decree No. 29/2003, Republic of Indonesia.

The CBM investors must be aware of this negative impact as well as many other impacts and provide appropriate technologies and solution to address this problem.

(v) **Noise**

A CBM project involves heavy equipment mobilisation. This mobilisation and the project activities will cause noise disturbance in the area

(Hulme 2002; Bryner 2002). This noise occurs during the entire life cycle of the project. Therefore, the CBM investors need to compensate affected communities due to the noise disturbance, to enable them to run the project smoothly.

5.3 Survey on the Local Community Impacts and Expectation Towards CBM Development

A survey was carried out to investigate the local impacts on four main CBM basins in which CBM operations have occurred.[1] The survey result showed that the locals are impacted both positively and negatively. The local impacts survey consisted of nine different impacts that were incurred in local communities: impact on employment; education; health care service; forestry; infrastructure; water quality; air, land and farms; and noise. A total of 130 questionnaires were gathered and analysed (Kutai Basin, 42 questionnaires; Barito Basin, 21 questionnaires; South Sumatra Basin, 40 questionnaires; and Central Sumatra Basin, 27 questionnaires).

- **Kutai Basin**
  The survey in Kutai Basin covered five different areas in Kutai Kartanegara District, but only three areas were analysed as they were located where the main CBM activities occurred: Sanga-Sanga, owned by VICO CBM; and Poros Kabo and Palaran, owned by Ephindo Energy. Besides the CBM activities, conventional oil and gas as well as coal mining activities took place in this basin, especially in Sanga-Sanga. Most of the local people in this area were unaware of the CBM activities since they were unsure about which activities actually impacted their area. In total, 42 questionnaires were carried out in Kutai Basin (Sanga-Sanga, 3; Poros Kabo, 10; and Palaran, 29). Only

[1] Based on author's survey in 2017–2018.

those questionnaires filled out by local people who really know about the CBM activities in this area were included.

- **Barito Basin**
  The survey in Barito Basin was carried out in one village, Walangkir, which was located in Tabalong district. A total of 21 questionnaires were obtained. The CBM block located in Walangkir belongs to PT Pertamina Hulu Energi (PT PHE), Metan Tanjung II. The local people were aware that this project was being carried out in their area and were thus able to give a reliable response to the questionnaires.
- **South Sumatra Basin**
  The areas visited in the South Sumatra Basin included the area in which CBM development has occurred, centred around three villages. This project is undergoing a review of its POD/FDP. The CBM well belongs to PT PHE. Currently, it is the only CBM project owned by NuEnergy Ltd. that is entering the POD approval process. Local people were aware of the CBM development being carried out in their area. A total of 14 questionnaires were obtained from Midangan Asri (the location of NuEnergy Ltds pilot CBM production wells), 12 questionnaires from Trans Sosial (NuEnergy), and 14 questionnaires from Sungai Tebu, a village close to the pilot CBM (PT PHW).
- **Central Sumatra Basin**
  Bukit Jaya Village (Riau, Pekanbaru) is located close NuEnergy Ltds CBM well. The well was located in the middle of the palm forest near to local housing. This well was being plugged and abandoned when the author visited the village since it was not very prospective technically. A total of 27 questionnaires were obtained during the survey.

The main issue according to Soetrisno (2018) when dealing with local communities and other parties who own the land (e.g. coal miners and palm companies) is the technical issues. The most important thing to be concerned about is the intangible cost of dealing with this:

> There are many uncertainties in this industry, not mainly internally, but in its technical side, is one of the main issues. But the most important thing is the unrecorded issues. (Soetrisno 2018)

The government should look at the big picture of the industry. They might not get much money from the project, but by having CBM activities in some areas can boost the economy in which the CBM activities are occurring:

> You see as previously said that the Government is thinking about how much money will be earned they are getting from the contractor? But if they look more, let's say big picture, in a big picture they don't realise that any promotion to the any industry, and we are talking about CBM now, will boost at least the provincial economic where these projects are taking place, and also will increase the number of place working for this project etc, and all this services will be required you know, that you have these projects. It will basically grow the economy in this provinces, despite that for the particularly upstream sectors, they want just much money for example right? I think that's why it's important mmm… this local is considered is very important. (Golubchenko 2018)

Local impact is considered an important factor that influences the CBM success in areas where they are located:

> I think it (local impacts) is important but probably the government doesn't even think it is important but the good thing is, both good and bad is that right now is not a significant contributor, the production just because the production right. (Utama and Golubchenko 2018)

As CBM gas is mostly used for domestic or local needs, there is no need to worry about global oil prices or gas prices:

> Mm but mostly CBM deliveries into domestic small scale network, like a village power or something like that, so you don't necessarily have to worry about global LNG prices or global oil prices. You just came up with the domestic contract that works. (Mair 2018)

In regards to local community involvement, the government assists the company to introduce their project and state what they can offer to the local community:

> The Government doing the standardized approach. The company and the government campaign socialization, community development, in cooperation with the related branch of Local government and community leaders. (Purba 2018)

The local participating interest was implemented 12 years after government law 35/2004 was enacted. There was no transparent implementation regarding the law, particularly for the local participating interest. Enforcement of MR 17/2016 has caused an issue for the investors as it has a direct influence on the economics of their projects:

> At the end, the government thinks about it, how, and at last the minister regulation was enacted, (MR37/2016), that obliged the contractors to give away their 10% of PI to BUMD (local Government institution), that is one of them. How the local people, that is expected by the local government institution, enjoy directly and even then, they should contribute to the cost as well, correct? 10% from the cost. But this BUMD has got no financial capability. Therefore in the MR it is stated, the contractor must bear the cost for the BUMD. So the BUMD really got a free 10%. (Husin 2018)

Gunawan (2018) believes that over the next ten years, CBM can become the key local energy source for local industry:

> The purpose, I think, maybe for the industry. The industry maybe for the electricity generation for local industry. In some areas, these are small industry, need energy sources to generate their electricity for their industry. Those that might not have been fulfilled by the PLN. I think in the next 10 years, I think they have not yet able to provide the electricity for the local industry.

There was a discussion on land acquisition with the vice minister, about the formalities, particularly for land acquisition. The idea was to get the land permit done collectively by SKKMigas:

> Lately, when there was an event by the Vice Minister at SKKMigas, we all gathered. What if for the formalities, most importantly the formalities, for

> the land acquisition – surely by the contractor- but for the permits, what if we do it collectively by the SKKMigas. There is regulation for this, but I forgot, where exactly, maybe in the agraria law or what. (Gunawan 2018)

(i) Local communities' perspective on CBM development

Figure 5.1 shows that only 1% did not agree on CBM development in their area due to the damage to water wells, and 4% were not sure what they could get from CBM development in their locality. The majority (95%) agreed and supported CBM development in their area with conditions (see Fig. 5.2).

In general, local communities agreed on having CBM development in their areas as long as they brought benefits to the society, which were classified into four main foci to be considered by the investors before commencing their projects. Most of them agreed on CBM development in their area as long as they help the communities developed economically. This can be seen from the percentages shown in Fig. 5.2, which include the improvement in social welfare and increasing local communities' income and job opportunities (76%). The other two conditions that they

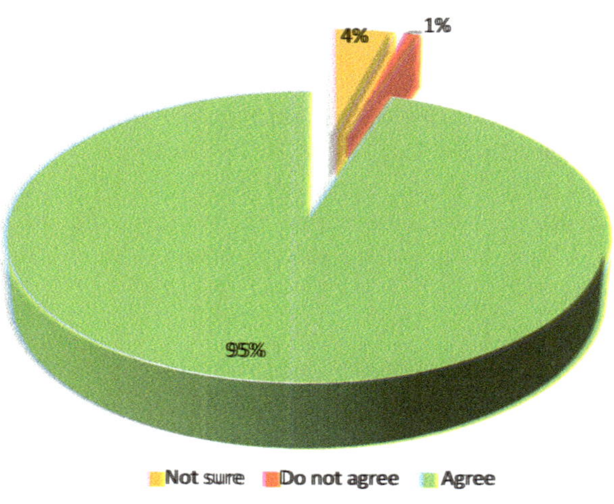

Fig. 5.1 Local communities' perspectives on CBM development. (Source: Created by the author (2020))

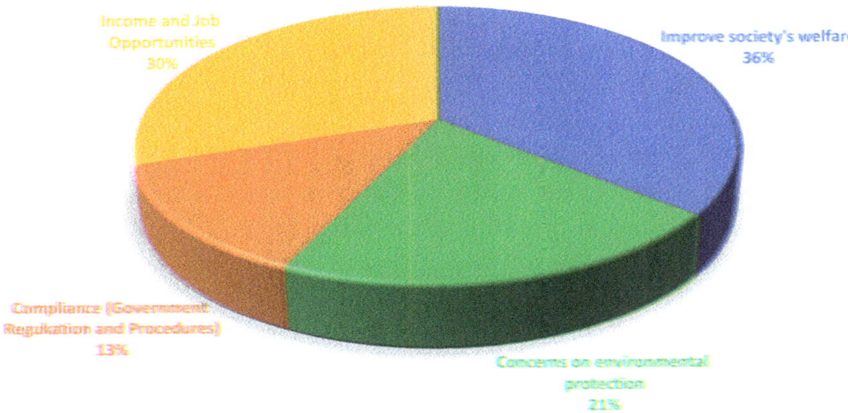

Fig. 5.2 CBM development requirement by the local communities. (Source: Created by the author (2020))

emphasised and that the investors should be concerned with included environmental impact and complying with any government regulations and procedures, including obtaining permits from the government before commencing the project.

The next part explains the comparative analysis of the local impacts and local expectations.

(ii) Comparative analysis on local impacts and local expectations

The comparative analysis here utilises two different comparative approaches: parameter based (local impacts parameter and local expectations parameter) and area based (the four different CBM basins). The parameter-based comparative analysis aims to understand the impacts and expectations in the four different areas and the reason why they have different levels of impacts and expectations. The area-based comparative analysis aims to understand that each area evidences different levels of impact, and this provides information to both GoI and investors regarding which sector needs to be prioritised. In terms of local expectation, this area-based comparative analysis provides information to both of GoI and investors which sector/expectation needs to be improved, developed or fulfilled during their project life cycle.

Moreover, these two different analyses will assist the GoI to impose a more appropriate fiscal regime in each area according to their unique situation and condition, both socially and geologically. Meanwhile, from the investors' point of view, this will help them to have a better plan and budget that is related to community development or CSR. This information will help the investors to avoid the issue of unpredictable costs related to the local communities. An agreed understanding amongst local communities, government and investors is very important for the continuity of the CBM projects.

5.3.1 Parameter-Based Comparative Analysis

The parameters being used in this analysis are the same parameters asked in the survey questionnaires. The parameters are employment, education, healthcare services, infrastructure, electricity supply, and environmental impacts. The environmental impacts consist of forestry, water quality, air, land and farms, and noise disturbance. The parameters of the local expectations are similar to the local impact parameters. The only difference is that the local expectation does not have the environmental expectation, but cash benefits. These analyses are discussed comprehensively below:

Employment

Employment here means the direct employment and indirect employment impacted by the CBM activities. Both are related to the income of the local people. Direct employment is employment directly by the CBM company to the local people, while indirect employment is employment that arose due to the existence of the CBM activities and brought a better income for the local people, such as shops, accommodation, and cleaning.

According to the survey, there was almost no impact on employment in the Kutai Basin, and the most impacted area was Barito Basin. Most of the local people in Sanga-Sanga (Kutai Basin) were unaware of the CBM activities in their area, with only three questionnaires taken into consideration. According to the locals in Kutai Basin, there was no significant impact from the CBM project on employment in the local communities, mainly due to the period of CBM drilling activities being done in their areas. There are many conventional oil and gas and mining activities in their areas which have provided more employment opportunities to the local communities. The CBM project in Kutai Basin did not last long due to financial and technical challenges (Husin 2018). The CBM investor only hired local people to clear the sites at the beginning of the project and one or two as security guards. In Sanga-Sanga (also located in Kutai Basin), with many mining and conventional oil and gas projects, more locals were hired than the CBM investors.[2] Only 17% responded that there was a good impact from the CBM development in their area, but 69% of the locals responded that the impact from CBM development was insignificant.

Different perspectives were given by the local people in ***Barito Basin, South Sumatra Basin,*** and ***Central Sumatra Basin***. The survey was done in an area in which the dewatering process was still occurring. It was obvious that the locals were very aware of the existence of CBM development, with more locals being employed as the CBM activities were still very active, especially in the Barito Basin where 90% of respondents stated good impacts from the existence of CBM projects. They mentioned that having this project in their areas meant that more locals were being employed as workers in the field. This project also improved their economy since their local shops had more customers.

However, the expectations of the locals in four main CBM basin areas were very high. Notably, employment was considered one of the most critical sectors in these areas. All of the four main CBM areas consider that it is very important that the CBM development in their area bring more job opportunities for local people. However, in Central Sumatra Basin (Bukit Jaya Village), only 48% consider that the increase in employment rate is very significant and 44% consider it important. Only 7% of them consider it to be unimportant. This is most likely because current economic situation in their community is developed enough and that

[2] This was seen by the author during the survey and confirmed by the locals.

most of their main income comes from the palm plantation as a result of the immigration which happened in the past. To encourage the spread of the population in the past, the GoI provided incentives to the most populated area (in Java Island) to migrate to this area by giving them 2 acres land for palm plantation, including housing.

In Kutai Basin, the local community expect to have better job opportunities as there is not enough impact on the local employment rate they have received as an impact of the CBM development in their area. This is similar to what is expected in the Barito Basin community, that having job opportunities for the local community is important. However, the author found in the survey process that the community lacked education and that most of them do not speak Bahasa Indonesia, but only their own dialect. In this case, they consider getting a better education is more important.

Meanwhile, in the South Sumatra Basin, the local community considers that it is significantly important to increase the employment rate in their area. During the survey, the author observed and found that this area is more developed and has more access to economic activities. However, a big mining company has opened many job opportunities for the local people. This has given a thought to the local people to have an expectation on the increase of employment rate as an impact of the CBM project development in their area.

Education

Education holds a significant role in the development of society, especially in a country like Indonesia, which has more than 17,000 islands (Geospatial Information Agency of Indonesia, 2017). A survey was conducted in the four CBM basins to see the impact of the CBM development towards education quality there.

There seems no impact on education in Kutai Basin, South Sumatra Basin, and Central Sumatra Basin. Most of the areas do not significantly benefit by having CBM development in their areas. Local people in Central Sumatra Basin admitted that there was no education impact from the CBM development. In South Sumatra Basin, 90% people also said that there was no impact on the education sector in their area, as well as the 76% of local people in Kutai Basin. Some factors may cause this. First, it is suspected that the local communities being surveyed are more developed; therefore, the education sector has enough support from the local government. The second reason is due to the number of investors operating there, and the CBM operations are not massive compared to other mining or conventional oil and gas operations. The third reason is due to the financial capability of the CBM investors. The CBM investors operating the CBM activities in which the areas were being surveyed are private CBM investors. Hamzah (2018) admitted during the FGD that they have run out of financial support since they spent most of their funds at the beginning, due to high costs and the complexity of the rules and regulations that did not allow them to use cheaper and more proper tools and equipment for CBM. Therefore, they did not spend money in non-technical matters. Another idea is that investors tend to avoid extra costs before reaching project commerciality.

Unlike the other three main CBM basins, the local community surveyed in the Barito basin has benefited from the CBM operations in their area. The state CBM company, PT PHE Metan Tanjung IV, operates the CBM project. This local area is not as developed as the other main CBM basins. Those local people who can speak Bahasa Indonesia are only those people being educated in the school. The cultures, rural development level, as well as the location contribute to how CBM investors take actions, especially where they were about to start operations. It is part of their strategy to get support from the local communities and to ease their operation. Therefore, the local communities in the Barito Basin are able to see the significant difference in their education sector after the CBM investors came in.

However, all of the local communities consider that education is essential to them, and they expect that the existence of the industry will contribute to the improvement in their education sector.

As mentioned earlier in the employment section, the local community in the area of Barito Basin consider education to be essential, and they expect improvement in the quality of education in their area. In general, all communities expect to have a better education in their areas as they are all aware of the importance of a good education.

Healthcare Services

One of the problems faced by rural communities and areas is the quality of healthcare services. Even though it is considered important, CBM companies seem to fail to improve the healthcare services in the development areas. The healthcare services in all areas seem not to be impacted by the existence of CBM development. Unfortunately, there were not many benefits that the local people received in the CBM exploration stage. They admitted that there was not much support from the industry for the healthcare service in their areas. In Barito Basin, only some local people admitted that they received help from the existing CBM investor in regards to their healthcare services.

It is evident that the local people expect to have a better healthcare service in their areas (Sumarno 2020). There are some expectations on the increasing of the number of doctors and frequency of doctor visits to the villages, increasing medicine and vitamin supplies for local perople, as well as providing a cheaper healthcare service for local people.

The local community in Barito Basin considered the health services as being very important. However, to some of them it was rather neutral, meaning that it will be better to have improvements in healthcare services. However, if there is no improvement at all, there is no big issue there as they thought they had received adequate service in healthcare services compared to other areas being surveyed. However, improvement in healthcare services is considered important for all of them.

Finally, from looking at the impact and expectation on healthcare services, it is clear that CBM investors should contribute more to the improvement in healthcare services as expected by the local communities. Currently, only a few locals received better healthcare services from the CBM investor.

Infrastructure

Infrastructure plays a significant role in the success story of the CBM project. The infrastructure being discussed here refers to those public facilities, such as public buildings, roads, and bridges. Infrastructure is important since this will help to improve the accessibility of healthcare services, education, and so on.

According to the survey results, local people in Kutai Basin admitted that the road was destroyed due to the heavy equipment used in the CBM activities. Only a few people said that some of the roads and other infrastructure were built and fixed by the CBM investors. This could be because there is more than one CBM investor in Kutai Basin, and moreover there are many other conventional oil and gas investors operating in Kutai Basin. Meanwhile, Barito Basin, South Sumatra Basin, and Central Sumatra Basin show that there were good impacts on their infrastructure, especially the road in their villages. The CBM investors developed and fixed the road for the communities.

According to the survey results, the improvement of infrastructure in the villages is considered important. These people expected to have better infrastructure in their village. The better the infrastructures in an area, the better quality of life the people will have. Accessibility and affordability are directly influenced by how developed the infrastructures in an area. The areas located near the CBM main basins are considered underdeveloped compared to the main island, Java, Indonesia. Access for food, goods, and services (health, education, etc.), as well as energy, have become more and more important. Infrastructure is fundamental needs for running the economic activities in any areas (Halstead and Deller 1997; Srinivasu and Rao 2013), and it will eventually lead to affordable prices on all their needs. Therefore, the communities expect to have more infrastructure development in their areas.

5.3.2 Environmental Impacts VS Cash Benefits

CBM activities are known to have direct impacts on the environment (Rodvelt 2014; Meng 2017; Mayer 2016). It damages the environment due to its operational activities, such as fracking. The survey was conducted to determine the impacts on forestry, water quality, air, land/farms and noise.

Based on the results of the survey, a number of the local people in Kutai Basin admitted to having negative impacts (46%) and positive impacts (52%) from the CBM activities. This may be because there are many industrial activities there, so they could not recognise where the impacts came from. In other words, the results may reflect the overall impacts rather than from the CBM activities alone. In the other three main CBM basins (Barito, South Sumatra, and Central Sumatra Basin), most of the local people admitted that they were not impacted by the CBM activities. There were very few CBM activities in those three basins, with only a few wells drilled, as this was the pilot project stage.

The local people were asked whether they expected financial compensation or any actions to create a better environment; such as replanting the forest after the project finished and maintaining and fixing the roads and infrastructure that were impacted. The local expectation on financial compensation in Kutai and Barito Basins were very high compared to South Sumatra Basin. Local people in Kutai Basin tended to have the highest expectation on cash benefits, and this is considered very important for them since they have the most significant population, and they are more competitive in business or having a job (See Table 5.1). In addition, the local people have got used to the condition, and they have become less caring about the environment.

The local people in the Barito Basin also expected to get cash benefits since they did not see any environmental damage in their area due to the CBM activities. Therefore, they prefer to receive financial compensation.

Table 5.1 Total Population in the survey areas

CBM basin	Local district	Population	Year	Source	Date accessed
South Sumatra Basin	Muara Enim	618,760	2017	Muaraenimkab.bps.go.id	18/06/19
Central Sumatra Basin	Pelalawan	417,498	2016	Riau.bps.go.id	18/06/19
Barito Basin	Tabalong	218,620	2010	Kalsel.bps.go.id	18/06/19
Kutai Basin	Kutai Kartanegara	717,789	2015	Kurkab.bps.go.id	18/06/19

Disclaimer: The year is not comparable due to limitation on the information provided by the Central Bureau of Statistics Republic of Indonesia
Source: Created by author based on government official websites (2019)

Meanwhile, in South Sumatra Basin, they expected cash benefits, but, for them, this is considered not very important. They admitted that if they get cash benefits, they are happy to receive, but this is not the most significant thing for them. It is a different case in Central Sumatra Basin: the local people in this area prefer to have a better and improved environment. The village that was visited and surveyed is called Bukit Jaya, where people from Java Island migrated years ago. The local people are transmigrants, and they have been given two acres of palm forest to earn their living. They have a good quality of life in this village; therefore, they considered that financial compensation was not very important, but that a better and improved environment was.

5.3.3 Electricity Current Supply Condition

One of the main aims of having a CBM industry in Indonesia is to potentially increase domestic electricity supply. As such, it is important to see the current condition on electricity supply in the local areas. Current conditions regarding electricity supply in all areas are very bad. The areas often have blackouts, and the people admitted that they expected to have a better electricity supply. Therefore, having CBM projects in their areas is expected to increase electricity supply in their areas.

Local people in Barito and Central Sumatra Basins admitted that the blackouts often happened in their areas. Some of them mentioned that it was not always because of poor electricity supply, but also due to bad weather. In Central Sumatra, there were blackouts and affordability issues in their areas. Meanwhile, Kutai and South Sumatra Basins are two areas in which many oil and gas activities have taken place for a long time, and these areas are more developed compared to the other two main Basins, Barito and Central Sumatra Basins. However, there were still power cuts every day in Kutai and South Sumatra Basins, even then it was not as bad as in Barito and Central Sumatra Basins.

Electricity is one of the basic needs in life; therefore, there is a high expectation that there will be additional electricity supply by having the CBM project in their areas. The results showed that when an area has a very bad electricity supply condition, they will see the additional supply of electricity as very important for them. The local people in Barito and Central Sumatra Basins considered that having a better electricity supply was important. Surprisingly, people in Kutai and South Sumatra basins considered the same, but that it was not as important as in Barito and Central Sumatra basins.

The local communities in South Sumatra Basin were the most neutral to this issue, with the second highest in Kutai Basin. These areas are more developed compared to the other two areas (Barito Basin and Central Sumatra Basin). The blackouts that happened in South Sumatra and Kutai Basin were within the acceptance level of the local people. However, the majority of the local people still considered it important to have a better supply of electricity in their areas.

5.3.4 Area-Based Comparative Analysis

Four different basins were surveyed: Kutai Basin, Barito Basin, South Sumatra Basin, and Central Sumatra Basin. As previously mentioned, this analysis provides information on what issues of concern focused on by the GoI and the investors related to CBM development in each area. This analysis also looks at the priorities set by both the GoI and investors related to the impacts and expectations of the local communities.

Kutai Basin

In most sectors, the locals did not notice whether there had been a good impact (benefits) or bad impact from the CBM development in their areas. Despite having environmental damages due to CBM project, the local societies tend to receive financial benefits or compensation for the damages instead of having the environment protected and recovered. This gives a signal that any investor needs to manage the society expectations and manage their CBM project's environmental impacts as well as a good strategy to influence the expectation.

Although there was damage mostly on roads in this area, observations by the author were unable to justify if it was only because of the CBM development or any other energy (oil and gas) development projects in that area. It was explained before that Kutai Basin is a resources-rich area (gas, oil, CBM, and coal). This was further compounded by the location of the CBM wells being very close to other oil and gas projects and coal mining projects. However, this impact still needs to be taken into consideration by the GoI and investors in their development activities.

In terms of fulfilling local expectations, there are three main sectors that the GoI and investors need to prioritise: healthcare services, job opportunities (employment), and education.

Regarding the healthcare services sector, there was almost no impact from the CBM development in the local areas according to the data. Therefore, it showed the highest expectation (88%) on having an improvement. Any CBM investors should be looking at this sector as part of their community development. Improvement on healthcare services are due to the high level of air pollution in that area due to many energy and mining projects that are happening there, and not just the CBM project.

As previously discussed in the parameter-based comparative analysis, a high expectation on employment level (86%) may be triggered by other projects that provide job opportunities for locals; therefore, they expected many more job opportunities from CBM development.

The expectation on having a better quality of education was the third highest (81%), and this can be due to the awareness of the local people on the importance of the education for their people.

Barito Basin

The area visited in this basin was very isolated even though there was a fairly good road to reach this village. However, the current electricity supply in this area was very bad. All of the respondents agreed that they currently did not have adequate electricity supply in their local communities. This means that the GoI and investors need to consider the issue of electricity supply.

The local communities also received a direct impact on employment from CBM development. The investor running the CBM project in this area is a National Company, PT Pertamina, and they have contributed to the education sector in that area. The society in this area want to have better education facilities, and the CBM investor has provided this to the community. As previously explained in the parameter-based analysis, the local communities struggled to speak the national language, Bahasa Indonesia, and they are aware that education is fundamental.

Without ignoring other expectations, the GoI and investors can create a strategy to fulfil the expectations, which in this case they should consider to have a community development programme mainly on the education sector, followed by healthcare services, employment, and infrastructure, and as a result of the CBM development, they could provide more electricity supply to the local communities.

South Sumatra Basin

According to Table 5.1, the population in the South Sumatra Basin surveyed is the second largest population after Kutai Basin. As previously discussed, this area is more developed than the Barito and Central Sumatra Basins. Apart from Kutai Basin, South Sumatra Basin is also a resource-rich area in oil, gas, and coal mining.

The results of the survey evidenced that the CBM projects benefited the local communities. It is important that investors should bring improvements to the socio-economy of the local communities. The local communities still have a very high expectation on job opportunities from the CBM development, which the investors need to be aware of and develop a strategy on hiring local people. This can also be supported by the GoI, for example, by imposing local content regulations in this area.

However, the lack of electricity supply and the blackouts provide information to the GoI and investors that having a CBM development should contribute to additional electricity supply for the local communities.

Regardless of its level of development, the local community thought that the CBM activities had brought both bad and good impact on the infrastructure. This has become the second highest expectation of the local people. Some of the area being surveyed did not have proper paved paths or roads which were also used for CBM development.

Central Sumatra Basin

Despite having a lot of land suitable for palm plantations, the local people in this area mentioned that they did not have a good supply of electricity, in terms of both affordable prices and the amount of supplies. This shows that the GoI should look at the electricity policy in rural areas like this area, and that the investor need to contribute to a better electricity supply and affordable electricity prices for local communities.

The survey results proved that the local communities were aware that financial compensation was not the most effective solution to the disturbances from CBM development in their area. This has made this area unique compared to the other three other areas. They were also aware of the development of their society and environment protection in their area. This means that the GoI and investors should prioritise what the local communities expected more, for example healthcare services, electricity supply, infrastructure development, and many other indirect benefits they can offer to the local communities.

These two comparative analyses have provided different information, and the methods and analyses can be a tool for both the GoI and investors to set their strategy on planning and budgeting for CSR activities. This could help reduce the risk of unpredictable social costs.

References

Bryner, G., 2002. *Coalbed Methane Development in the Intermountain West: Primer*. Colorado: University of Colorado Law School.

Cully, M., 2015. *Review of the Socioeconomic Impacts of Coal Seam Gas in Queensland*. Canberra: Australian Government, Department of Industry, Innovation, and Science.

Emam, E., 2015. Gas Flaring in Industry: An Overview. *Petroleum & Coal*, [online] Available at: http://www.vurup.sk/petroleum-coal [Accessed 17 October 2017].

Fisher, J., 2016. *Environmental Issues and Challenges in Coai Bed Methane Production*. [online] Available at: https://pdfs.semanticscholar.org/8f1d/347a247406833104535 3a6576656b42a8bc4.pdf [Accessed 8 June 2017].

Geospatial Information Agency of Indonesia, 2017. *Beranda*. [online] Big.go.id. Available at: <https://www.big.go.id/> [Accessed 26 September 2020].
Golubchenko, N., 2018. *Views on CBM Industry in Indonesia (Global Consultancy's Perspective)*.
Gunawan, B., 2018. *Views on CBM Industry in Indonesia (Government's Perspective)*.
Halstead, J. and Deller, S., 1997. Public Infrastructure in Economic Development and Growth: Evidence from Rural Manufacturers. *Community Development Society. Journal*, 28(2), pp. 149–169.
Hamzah, S., 2018. *A Focus Group Discussion on CBM Industry in Indonesia*.
Hartiniati, H., 2016. DAMPAK LINGKUNGAN DAN SOSIAL DARI PENGEMBANGAN CBM DI INDONESIA. *Jurnal Teknologi Lingkungan*, 12(2), p. 207.
Hulme, D., 2002. *Coalbed Methane Development in Wyoming's Powder River Basin*. [ebook] Colorado: University of Colorado Law School. Available at: http://scholar.law.colorado.edu/coalbed-methane-development-intermountain-west/4 [Accessed 19 March 2020].
Husen, D., 2008. *ION EXCHANGE & DEIONIZATION: TREATING COAL BED METHANE PRODUCED WATER*. [online] www.waterworld.com. Available at: https://www.waterworld.com/articles/iww/print/volume-8/issue-5/features/ion-exchange-amp-deionization-treating-coal-bed-methane-produced-water.html [Accessed 6 December 2018].
Husin, M., 2018. *Views on CBM Industry in Indonesia (Investor's Perspective)*.
Letts, L., 2012. *Coal Seam Gas Production – Friend Or Foe of Queensland's Water Resources?*. 1st ed. Australia: Thomson Reuters.
MacKinnon, A. and Fox, K., 2006. Demanding Beneficial Use: Opportunities and Obligations for Wyoming Regulators in Coalbed Methane. *Wyoming Law Review*, [online] 6(2), pp. 369–400. Available at: https://scholarship.law.uwyo.edu/cgi/viewcontent.cgi?article=1135&context=wlr [Accessed 19 March 2020].
Mair, D., 2018. *Views on CBM Industry in Indonesia (Global Consultancy's Perspective)*.
Mayer, A., 2016. Risk and benefits in a fracking boom: Evidence from Colorado. *The Extractive Industries and Society*, 3(3), pp. 744–753.
Meng, Q., 2017. The impacts of fracking on the environment: A total environmental study paradigm. *Science of the Total Environment*, 580, pp. 953–957.

Mudd, M., 2012. Montana v. Wyoming: An Opportunity to Right the Course for Coalbed Methane Development and Prior Appropriation. *Golden Gate University Environmental Law Journal*, 5(2), pp. 297–342.

Purba, S., 2018. *Views on CBM Industry in Indonesia (Government Perspective , Written Interview).*

Rodvelt, G., 2014. Vertical Well Construction and Hydraulic Fracturing for CBM Completions. *Coal Bed Methane*, pp. 101–135.

Setyatmoko, U., 2018. *Views on CBM Industry in Indonesia (Investor's Perspective).*

Skov, J. and Myers, N., 2004. *Easy Money, Hidden Costs: Applying Precautionary Economic Analysis to Coalbed Methane in the Powder River Basin.* [online] Iowa: Science and Environmental Health Network. Available at: http://www.sehn.org/tccpdf/coalbed%20methane%20costs%20Powder%20River.pdf [Accessed 16 September 2017].

Soetrisno, S., 2018. *Views on CBM Industry in Indonesia (Industry Observer's Perspective).*

Srinivasu, B. and Rao, P., 2013. Infrastructure Development and Economic growth: Prospects and Perspective. *Journal of Business Management & Social Sciences Research (JBM&SSR)*, 2(1), pp. 81–91.

Sumarno, T., 2020. *Identifying and Modifying the Coal Bed Methane Fiscal Regime in Indonesia: A Comparative Study with the Australia and the U.S.. Doctorate*. University of Dundee.

The Pembina Institute, 2003. *The Environmental Challenges of Coalbed Methane Development in Alberta*. [online] Drayton Valley: The Pembina Institute. Available at: https://www.pembina.org/reports/CBM_Final_April2006D.pdf [Accessed 11 February 2020].

Utama, J. and Golubchenko, N., 2018. *Views on CBM Industry in Indonesia (Global Consultancy's Perspective).*

Waeckerlin, E., 2005. The New Border War: CBM Development in the Powder River Basin of Wyoming and Montana. *Public Land & Resources Law Review*, 26, pp. 149–162.

6

Reattracting CBM Investment in Indonesia

Abstract This chapter provides ideas on how countries (with a focus on Indonesia as the case study) can achieve energy transitions goals by reattracting CBM investment. How would the government be expected to be involved in the success of CBM development and support investors to provide equitable benefits to society? Issues are defined earlier (in previous chapters), and this chapter highlights what can be done to solve the issues concerning society and environmental justice through taxation by utilising examples from the US and Australia. Taxation is one of the key tools in the development of any country and is very much related to welfare distribution; therefore, this chapter highlights taxation modifications to improve CBM development and create project attractiveness for governments and investors. Lessons from other countries are drawn from this chapter in the process of suggesting modifications to the taxation system in Indonesia.

Keywords Production-sharing contract • Petroleum Resource Rent Tax • Royalty/tax • Severance tax • Ad-valorem tax • Indonesia • Australia • Wyoming • Fiscal regime • Fiscal incentives • Taxation • Government shares • Uplift rate • The US

T. B. Sumarno, *Just Energy Transitions and Coal Bed Methane*, Energy, Climate and the Environment, https://doi.org/10.1007/978-3-030-85490-4_6

6.1 Introduction

This chapter provides ideas on how countries (with a focus on Indonesia as the case study) can achieve energy transition goals by reattracting CBM investment. How should the government be expected to be involved in the success of CBM development and support investors to provide equitable benefits for society? This chapter also highlights what can be done to solve the issues concerning energy justice through taxation, utilising examples from the US and Australia. Taxation is one of the key tools in any country's development and is very much related to welfare distribution. Therefore, this chapter highlights the taxation modifications needed to incentivise and improve the potential for CBM development and create confidence in these projects for both government and investors. The purpose of this analysis is to better understand comparatively how other countries that have been successful in developing CBM deal with the industry. This chapter determines the fiscal regimes from these two countries that could be adapted in the newly constructed fiscal regime for Indonesia's CBM, including the Net PSC from Indonesia, the royalty/tax (R/T) from the US, and the Petroleum Resource Rent Tax from Australia.

6.2 Indonesian CBM Fiscal Regime: Net PSC

This PSC follows and complies with the regulations attached to PSC applied to conventional hydrocarbons. Table 6.1 sets out the fiscal terms in the PSC applied to CBM in Indonesia.

6.3 The US CBM Fiscal Regime: Wyoming State Fiscal Regime

The fiscal terms being applied are attached to which state the minerals/petroleum resources are located. The Powder River Basin is located in both states of Wyoming and Montana, an area of approximately 25,800 square miles, of which 75% is located in Wyoming (All Consulting

Table 6.1 PSC fiscal terms

Fiscal terms	Description
Gross production	Total production of a project every year before deducted by any expenditures.
Gross revenue	Based on total lifting multiplied by the standard price which is being used in Indonesia.
First Tranche Petroleum (FTP)	It is conceptually royalty. However, Indonesia tends to avoid the use of the term 'royalty'. The idea of the FTP is to secure the GoI shares before everything else is being imposed. Before 2010, the percentage was 10% and non-shareable between the GoI and investor. After 2010, the FTP has increased to 20% and shareable between the GoI and investor according to their agreed split. It is a front-end loaded fiscal term for investor.
Investment credit (IC)	An incentive given by the GoI to investors to recover its capital expenditures. This is an additional cost recovered during specific periods and it is taxable (Tordo 2007).
Marginal field incentive	It is another type of investment credit, which is also an additional cost to be recovered. This is an incentive given by the government to the investor for the working area, which has rate of return below 15%. The additional cost to be recovered is 20%. This is a non-taxable item (PriceWaterhouseCoopers 2015).
Sunk cost	All expenditures occur before the production stage.
Cost recovery (CR)	Exploration (sunk costs) and production expenditures being recovered by the net production after deducted by the FTP. There is no limit on CR, and the FTP has actually the cap of the CR, since the CR can only be as high as 90% (before 2010) or 80% (after 2010) of the total production. (gross production—FTP)
Domestic market obligation (DMO)	This is an obligation for the contractor to sell the oil production to fulfil the domestic needs. The DMO fees is a price agreed between the GoI and the investor in the contract.
DMO holiday	A type of incentive granted by the GoI to investors by buying the agreed production for domestic needs in a market price, and not a price set as the DMO fees.
Ring-fence policy	It is a policy on how the GoI allows the investor to merge their costs from different block/field. This has significant impact on the project economy, as when it is applied to fields, investors cannot merge the exploration cost incurred in an uneconomic field to the production field. The sunk costs will solely be borne by the investor, yet they have to pay a higher tax in the production field. On the contrary, the GoI obtains a greater benefit by applying this as it enables the GoI to collect the income tax earlier.

Source: Summarised by the author (2020)

2004); therefore, this research will look closely at the fiscal system applied there.

According to Jackson (n.d.), Wyoming repealed the tax severance credit applied to CBM in 2000 since they realised that the CBM was booming and growing rapidly. Jackson (n.d.) wrote that CBM originating from wells drilled after 1 April 2000 are subject to the same state severance tax rate as traditional natural gas producers.

Wyoming has applied severance and ad valorem taxes (county gross products tax), while Montana only applied severance tax to their mineral and petroleum activities. According to the Council of State Governments (2016), severance tax in Wyoming is defined as an excise tax imposed on the present and continuing privilege of removing, extracting, severing or producing any mineral in this state.

This part also focuses on the Wyoming taxation system because it is a more comprehensive system compared to Montana. Therefore, a similar understanding will be gained regarding how the fiscal system is applied in Montana.

6.3.1 Taxation System in Wyoming

The Wyoming Statutes (W.S.) stipulate the terminology used in the taxation system in the Wyoming Constitutions. According to the W.S. 39-14-101 (xv), "Value of the gross product" means fair market value less any deductions and exemptions allowed by Wyoming law or rules. This shall be the fair market value of the product of the mine on the month after the mining or production process is completed.

Johnston (2019) found the nomenclature for the terms used in the Wyoming taxation system to be unclear. Hence, before further discussing the Wyoming fiscal structure, Table 6.2 describes the terms that will be used in this book to avoid any misunderstanding. The terms are taken from both the Wyoming legislature website (the Wyoming State Legislature, accessed in 2019, https://www.wyoleg.gov)[1] and Analysis

[1] The Wyoming Legislature (https://www.wyoleg.gov/InterimCommittee/2015/SMTPresentation AG-DoR.pdf and https://www.wyoleg.gov/InterimCommittee/2018/09-201811293-07Wyoming ConceptsonOilandGas.pdf)

Table 6.2 Terms and definition for Wyoming taxation system

Fiscal terms	Description
Netback price/value	The value of the gross revenue at the POS deducted by all costs associated with the transporting the gas from POV to POS.
Royalty	There are two main types of royalty in Wyoming, federal royalty and private royalty. The federal royalty is payable to the federal government if resource is located in federal land; therefore, this type of royalty is deductible for the purpose of calculating the severance and the property tax (Daniel Johnston & Co., Inc. 2018). Meanwhile, the private royalty is payable to the landowner, and this type of royalty is not exempt in Wyoming (The Wyoming Legislature 2019). Another type of royalty is state royalty, which is imposed directly to resources located in state's land. The state royalty is never shared with the federal, but if the resources is in a land belongs to federal government, federal royalty is imposed directly to the production, but part of the royalty belongs to the state (The Wyoming Legislature 2019; Daniel Johnston & Co., Inc. 2018). The location of the resources is not significant for the contractor, but it is important for the government as wherever the resources is located, the contractor pays royalty either to the federal (shared with the state government) or to the state government. These royalties are front-end loaded with netback value as the royalty based. Royalty provides government revenue at the earliest stage after all the allowable deductions for the federal and state royalties.
Taxability ratio	The ratio of taxable value to total sales value (www.wyoleg.gov)
Taxable value	The taxable value is determined by subtracting allowable transportation deductions, processing deductions and exempt royalties (Daniel Johnston & Co., Inc. 2018).

(*continued*)

Table 6.2 (continued)

Fiscal terms	Description
Ad valorem taxes and severance taxes	The ad valorem taxes are also called property tax, county gross products tax, or mill tax. Severance taxes are basically taxes on production, meaning that these are imposed only when the gas is produced, while the ad valorem taxes in Wyoming is based on the previous-year production (The Wyoming Legislature 2019). The ad valorem taxes are the creation of Wyoming Constitution, and therefore are bound by the constitutional constraints, while the severance tax is a creation of the statues. The tax base for calculating both of these ad valorem and severance taxes is the netback value deducted by royalties (Daniel Johnston & Co., Inc. 2018). Both of these taxes are payable to the state government which have become one of the state government revenue sources, but not for the federal government. These taxes are imposed before the capital expenditures and operating expenditures (front-end loaded), which makes the system regressive and which provides guarantee for the state government to receive their shares even when the project is not profitable. This seems to not be very favourable towards the contractor, if the project is marginal. However, these taxes also reduce the federal government revenue from the taxable income, CIT.
Corporate income tax (CIT)	The CIT is imposed after all deductions (royalties, the state government taxes, CAPEX and OPEX of the project). This is payable to the federal government, but not to the state government. This CIT is back-end loaded since it is only imposed at the end, after all the deduction.

Source: Summarised by the author (2020)

and Commentary of Wyoming Legal and Fiscal Frameworks by Daniel Johnston & Co., Inc. (2018).

Daniel Johnston & Co., Inc. (2018) wrote clearly about the Wyoming fiscal structure. Wyoming State has three main systems in its fiscal regime: federal, state, and private or fee acreage. This system is mainly constructed by two main types of the fiscal element:

1. *Revenue-based mechanisms* such as royalties (federal or state), severance taxes, and property taxes (or ad valorem taxes or gross product taxes).
2. *Profits-based mechanisms* such as state and federal income taxes.

In their 2017 report on Oil, Gas, and Coal Royalties, the Government Accountability Office (GAO) stated that the State of Wyoming royalty can vary from 12.5% to 16.67%, depending on demand.

Federal royalty impacts on the cash flow for both the state's take (revenue) and the ad valorem tax, which then also impacts on the taxability ratio. The increases on federal royalty are in line with the state's revenue but not with the taxability ratio and the ad valorem tax. The higher the federal royalty rate, the lower is the taxability ratio and the ad valorem tax (The Wyoming Legislature 2019).

Table 6.3 illustrates how the fiscal structure applied in Wyoming with the sales price at the point of sale (POS) is \$3.00/mcf with \$1.00 netback to point of valuation (POV) for the purpose of calculating the royalty and other taxes such as conservation tax, environmental tax and other levies set at the 0.8% rate (Daniel Johnston & Co., Inc. 2018). However, the royalty rates are using the rates given by Wyoming state's government (The Wyoming Legislature 2019). The federal royalty rate is 12.5%, with the allocation to the state government of 48%. The severance tax and ad valorem tax rates are 6% and 6.9%, respectively, according to the Wyoming Oil and Gas State Taxes official document, signed and authenticated by the Bureau of Land Management, United States Department of Interior (2018). The ad valorem taxes in Wyoming very much depend on the county and are based on mill rates; therefore, they can vary in each county.

Table 6.3 shows how the federal royalty is impacting on the ad valorem tax, state revenue and the taxability ratio. The calculations start with gross

Table 6.3 Federal royalty implications

	Terms	Ex. 1	Ex. 2
A	Total revenue	$1,00,00,000.00	$1,00,00,000.00
B	Allowable deduction for determining federal royalties	$20,00,000.00	$20,00,000.00
C = A-B	**Royalty bearing value**	**$80,00,000.00**	**$80,00,000.00**
D	Percentage of acreage federally owned	100%	50%
E	Federal royalty rate	12.50%	12.50%
F=C*D*E	**Federal royalty owed**	**$10,00,000.00**	**$5,00,000.00**
G	Allowable deduction for determining Wyoming production taxes	$20,00,000.00	$20,00,000.00
H=A-F-G	**Wyoming production tax taxable value**	**$70,00,000.00**	**$75,00,000.00**
I	Wyoming allocation of federal royalty	48%	48%
J=F*I	**Wyoming state dollars of federal royalty**	**$4,80,000.00**	**$2,40,000.00**
K	Wyoming severance tax	6%	6%
L=H*K	**Wyoming severance tax collected**	**$4,20,000.00**	**$4,50,000.00**
M=J+L	**Total tax and royalty to Wyoming state**	**$9,00,000.00**	**$6,90,000.00**
N	Ad valorem tax rate from gas producing tax districts	6.90%	6.90%
O=H*N	**Ad valorem tax collected**	**$4,83,000.00**	**$5,17,500.00**
P= M+O	**Total state and county**	**$13,83,000.00**	**$12,07,500.00**
Q=H/A	**Taxability ratio**	**70%**	**75%**

Source: The Wyoming legislature (2019)

revenue at the point of valuation. According to W.S. 39-14-203 (b) (iv), the natural gas POV is when the production process for natural gas is completed:

> The production process for natural gas is completed after extracting from the well, gathering, separating, injecting and any other activity which occurs before the outlet of the initial dehydrator. When no dehydration is performed, other than within a processing facility, the production process is completed at the inlet to the initial transportation related compressor, custody transfer meter or processing facility, whichever occurs first.

The gross revenue value is equal to the product price at the fair market price at the point of sale, and the royalty-based value which is also called the netback price/value is the gross revenue value less transportation costs, allowable deductions which are processing deductions and exemptions (e.g. public property). All costs associated with transporting the gas from POV to POS are included in the transportation costs.

The following part will discuss the economics of gas project in Wyoming in which the PRB is mainly located. It will clearly explain how the federal government, state government, and investors get their share.

6.3.2 Economics Structures of Fiscal System in Wyoming

Having discussed the fundamental concept and understanding on how the fiscal structure in general works, this section will provide further understanding on the implementation of the system and on the shares between the government and the investor.

As previously mentioned, there are three types of different royalties in Wyoming (federal royalty, state royalty and private royalty). This depends on the ownership of the land in which the resources are located, a factor that slightly influences the structure in the economic evaluation in determining the government's share.

The PRB is located in the State of Wyoming; however, this section will also discuss and explain further on the implementation of the R/T system applied to the resources located on federal land to see the differences and get a better understanding on the impact of the taxes imposed to the project.

Table 6.4 sets out an example of 1 mcf gas on federal land assuming that the sales price at the POS is $3.00/mcf with $1.00 netback to the POV for royalty determination purposes. The state's revenues come from state taxes, which are severance tax, ad valorem tax, and other taxes (which includes conservation tax, environmental tax and other miscellaneous levies) as well as 48% of the federal royalty. There is no income tax imposed by the state as it is imposed at the federal level, meaning that this is a tax income for the federal government. The federal government also receives a 12.5% royalty payment from the company less the 48% share to the state. The government take consists of the federal government's profits share and state government's profits share.

Table 6.4 Economics of 1 mcf at US$2/mcf of gas on federal land

	US$		Remark	
Gross revenue	$2.00		*Netback value*	
Federal royalty	-$0.25		*12.50%*	*(48% goes to the state)*
Net revenues	$1.75			
Severance tax	-$0.11		*6%*	
Ad valorem tax	-$0.12		*6.90%*	
Other tax	-$0.01		*0.80%*	
Net revenues	$1.51			
Capital expenditures (CAPEX)	-$0.42			
Operating expenditures (OPEX)	-$0.30		*Lease operating expenses*	
Total costs		-$0.72		
State taxable income	$0.79			
State income tax	-$0.00			
Federal tax base	$0.79			
Federal tax	-$0.17		*22% tax rate*	
Company's cash flow	$0.62			
Division of revenues ($ and %)				
Federal share	$0.30	15.2%	*100%–48% of federal royalty + federal tax*	
State share	$0.36	18.0%	*48% + severance tax +ad valorem tax + other tax*	
Company share	$1.34	66.8%	*Investor's cash flow + total costs*	
Total revenues	**$2.00**			
Total costs	**$0.72**	36.0%	*of gross revenues*	
Economic profits	**$1.28**	64.0%	*of gross revenues*	
Division of profits ($ and %)				
Federal share	$0.30	23.7%		
State share	$0.36	28.1%		
Company share	$0.62	48.2%		
	$1.28			
Government take	**51.8%**		*State + federal (profit shares)*	
Company take	**48.2%**			

Source: Daniel Johnston Inc. (2018)

Table 6.5 is an example of the economics on state land using the same data and assumptions in Table 6.4. The distinction between them is the federal royalty and the state royalty. As explained, the royalty is paid directly to the owner of the land. Table 6.5 shows the economics of a gas

Table 6.5 Economics of 1 mcf at US$2/mcf of gas on state land

	US$				*Remark*
Gross revenue	$2.00		$3.00		*Netback value*
State royalty	-$0.33		-$0.50		*16.67%*
Net revenues	$1.67		$2.50		
Severance tax	-$0.10		-$0.15		*6%*
Ad valorem tax	-$0.11		-$0.17		*6.90%*
Other tax	-$0.01		-$0.02		*0.80%*
Net revenues	$1.44		$2.16		
Capital expenditures (CAPEX)	-$0.42		-$0.42		
Operating expenditures (OPEX)	-$0.30		-$0.30		*Lease operating expenses*
Total costs		-$0.72		-$0.72	
State taxable income	$0.72		$1.44		
State Income Tax	-$0.00		-$0.00		
Federal tax base	$0.72		$1.44		
Federal tax	-$0.16		-$0.32		*22% tax rate*
Company's cash flow	$0.56		$1.12		
Division of revenues ($ and %)					
Federal share	$0.16	7.9%	$0.32		*100%–48% of federal royalty + federal tax*
State share	$0.56	28.1%	$0.84		*48% + severance tax +ad valorem tax + other tax*
Company share	$1.28	64.0%	$1.84		*Investor's cash flow + total costs*
Total revenues	**$2.00**		**$3.00**		
Total costs	**$0.72**	36.0%	**$0.72**	24.0%	*of gross revenues*
Economic profits	**$1.28**	64.0%	**$2.28**	76.0%	*of gross revenues*
Division of profits ($ and %)					
Federal share	$0.16	12.3%	$0.32	13.9%	
State share	$0.56	43.9%	$0.84	37.0%	
Company share	$0.56	43.8%	$1.12	49.2%	
	$1.28		$2.28		
Government take	**56.2%**		**50.8%**		*State + federal (profit shares)*
Company take	**43.8%**		**86.1%**		

Source: Daniel Johnston Inc. (2018)

project in state land, which requires the company to pay a royalty directly to the state government rather than to the federal government. The rest of the mechanisms are the same.

These tables show a different level of government and company take. Given the same cost and tax rates, the government take of the resources on state land is higher than on federal land. This is clearly due to the higher royalty rates in the state land (16.67%) compared to the federal land (12.5%). Assuming all costs are the same, but with higher revenue, the government take is actually lower, and this is due to the US government relying so much on the royalty for its revenue, and royalty or royalty equivalent fiscal elements are regressive. The higher the revenue (*POS at $4.00/mcf with $1.00 netback to the POV*), the lower the government take is (see Table 6.6). 'Royalty' is a fiscal term which has a regressive characteristic which can be managed by modifying the royalty types to fit into both the government's and the investors' interests.

This section has explained the royalty/tax system in Wyoming and how it impacts the government's take and investor's take. The system in the US is easy to administer yet regressive, but it gives investors the flexibility to manage their project, both technically and financially. The next section discusses the Australian fiscal system as applied to their CBM. It is slightly different to the R/T mechanism, and PRRT is what distinguishes the system in Australia from that of the US.

6.4 Australian CBM Fiscal Regime (Petroleum Resources Rent Tax)[2]

The Australian fiscal regime is a combination of CIT, PRRT, and royalty-based taxation. Table 6.7 provides a brief explanation on the specific fiscal terms in the PRRT system.

---

[2] Prior to the Treasury Amendment that took effect on 1 July 2019, Treasury Laws Amendment (2019 Petroleum Resource Rent Tax Reforms No. 1) Act 2019. This act removed onshore coal seam gas from the PRRT regime. From 1/7/19, a modified PRRT regime now only applies to offshore petroleum in Australian waters.

Table 6.6 Economics of 1 mcf at US$3/mcf of gas on state land

	US$		*Remark*
Gross revenue	$3.00		*Netback value*
State royalty	-$0.50		*16.67%*
Net revenues	$2.50		
Severance tax	-$0.15		*6%*
Ad valorem tax	-$0.17		*6.90%*
Other tax	-$0.02		*0.80%*
Net revenues	$2.16		
Capital expenditures (CAPEX)	-$0.42		
Operating expenditures (OPEX)	-$0.30		*Lease operating expenses*
Total costs		-$0.72	
State taxable income	$1.44		
State income tax	-$0.00		
Federal tax base	$1.44		
Federal tax	-$0.32		*22% tax rate*
Company's cash flow	$1.12		
Division of revenues ($ and %)			
Federal share	$0.32	10.5%	*100%−48% of federal royalty + federal tax*
State share	$0.84	28.1%	*48% + severance tax +ad Valorem tax + other tax*
Company share	$1.84	61.4%	*Investor's cash flow + total costs*
Total revenues	**$3.00**		
Total costs	**$0.72**	24.0%	*of gross revenues*
Economic profits	**$2.28**	76.0%	*of gross revenues*
Division of profits ($ and %)			
Federal share	$0.32	13.9%	
State share	$0.84	37.0%	
Company share	$1.12	49.2%	
	$2.28		
Government take	**50.8%**		*State + federal (profit shares)*
Company take	**49.2%**		

Source: Daniel Johnston Inc. (2018)

6.4.1 Economics Structures of Fiscal System in Australia (Onshore Projects)

This section demonstrates the flow of the combined fiscal regime in Australia. The PRRT liability was calculated based on the notional taxable profits. The deductions for the purpose of PRRT calculations were

Table 6.7 Fiscal terms—Australian PRRT: No longer applies to onshore projects from 1/7/19

Fiscal terms	Description
Assessable receipts	Assessable receipts included ordinary and statutory income, such as petroleum revenue, capital gain, tolling receipts, property receipts, employee amenity receipts, miscellaneous receipts, exploration recovery receipts, and incidental production receipts.
State royalty	State royalty is still imposed on an onshore project with a fixed percentage of 10% to 12.5% and paid to the state government. The base for royalty was the wellhead value. This royalty was deductible against PRRT liability. This royalty was imposed before the PRRT and only payable to the state government, which was deductible against PRRT liability, which was payable to the federal government. Despite being a front-end loaded for the system, this royalty did not actually make the system progressive as it was deductible against PRRT, which was payable after all deduction (CAPEX and OPEX). The rate of the royalty is reasonably significant for the contractor.
Wellhead value	Wellhead value was the petroleum revenue after all the deductions (transportation cost, excise payments, etc.) to get to the first point of sales (in the R/T system in the US; this was called the netback value/price).
PRRT	PRRT is a profit-based tax which was levied on onshore petroleum projects. The PRRT rate was set at 40%, which was imposed on onshore petroleum projects. PRRT payment was deductible against income tax. PRRT was imposed on a project basis and had applied ring-fence policy on a project base, but it allowed to transfer the project's exploration expenditures when they made excess on the assessable receipts after all deductions associated to that project. This was thought to encourage more exploration by the contractor, knowing that whenever they had PRRT taxable profit, these other exploration expenditures would be more likely to be a deduction. This was an incentive given by the government to the contractor, but this mechanism postponed and reduced the government revenue from the PRRT. In the case of royalty payments to the state government, the amount of the PRRT liability was reduced by the deducted royalty payment to the state government. The rate of the royalty would impact the federal government income; the higher the state royalty was, the lower was the PRRT liability to the federal government and the other way around.

(*continued*)

Table 6.7 (continued)

Fiscal terms	Description
Transferability for exploration costs and reduced uplift	Regardless of the ring-fence policy applied in the PRRT system (ring-fence policy only prevented the transferral of CAPEX and OPEX expenditures), transferring the exploration expenditures between projects was allowed. Transferring exploration costs within the same company group were to encourage and incentivise the industry to conduct exploration activities. This means that the uplifted green fields exploration costs could ultimately be offset against the assessable receipt from different projects within the group in Australia.
Taxing point	The first point at which a product became marketable.
Loss carry forward	The excess of deductions against total receipts in any year was compounded forward at long-term bond rate (LTBR) plus (uplift rate) (for long-term bond rate, see the following explanation on uplift rate).
Uplift rate	The uplift rate was an incentive given to investor due to the loss they suffer as the expenditures (exploration, development, and production costs) exceed the assessable receipts. The rate (LTBR plus) had varied between 5%, 15%, and GDP deflator. Uplift rates were significant for projects with high development costs and long lead time before commencing their production. The purpose of the uplift rate was very general and was not intended for specific purposes, and rather, it was to cover other ineligible expenditures such as interest on a loan. However, having this uplift rate compensated the loss of the value of money due to unrecovered costs in particular years.
Long-term bond rate	The government published rate applied to the loss carried forward for exploration, CAPEX and OPEX.[a]
Incentives on frontier exploration	The incentive was introduced in 2004/2005 as part of the Commonwealth budget. This incentive provided 150% allowance for a designated offshore frontier project for five years on its PRRT project. This was not applicable to CBM projects since it was onshore.

(continued)

Table 6.7 (continued)

Fiscal terms	Description
Closing-down expenditure	Closing-down expenditure was a recognised category of deductions. Any expenditures related to closing down of a petroleum project was categorised and deducted as closing-down expenditures. This was not transferable and could not be carried forward. This expenditure was realised in the year it incurred and therefore might attract tax credit. This usually happened when there was no longer production or income generation. This closing-down expenditure was attached to the local community and environment safety and sustainability. Therefore, the government and contractor should have encouraged each other, and the tax credit would encourage the contractor to do the closing (ASR) properly. It would not provide additional profits for the investors, but it could credit the tax they paid during their profitable years.
Tax credit	A 40% tax credit is still available to offshore projects that have previously paid PRRT for the excess of the closing-down expenditures as there were not enough assessable receipts for the related year. However, this credit cannot exceed the PRRT that has been paid.
Federal income tax (FIT)	The FIT is referred to as company income tax (CIT) in Australia. The CIT rate is 30% of taxable income (net of PRRT liability deduction).

Source: Summarised by the Author (2020)

[a]https://www.ato.gov.au/Rates/PRRT-augmentation-and-GDP-factor-rates/

royalty payment, general expenditures (including CAPEX, OPEX, and closing-down expenditures) during the exploration period, project exploration expenditures, and other areas' exploration expenditures (when the notional taxable profits are incurred). In this case (State of Queensland), the royalty was payable to the state government while PRRT and CIT were payable to the federal government. Figure 6.1 illustrates the flow of the fiscal system in Australia for resources located in the state jurisdiction.

Using the same number from the PRRT base calculation, Table 6.8 sets out the overall calculation to see the government's shares and the contractor's shares.

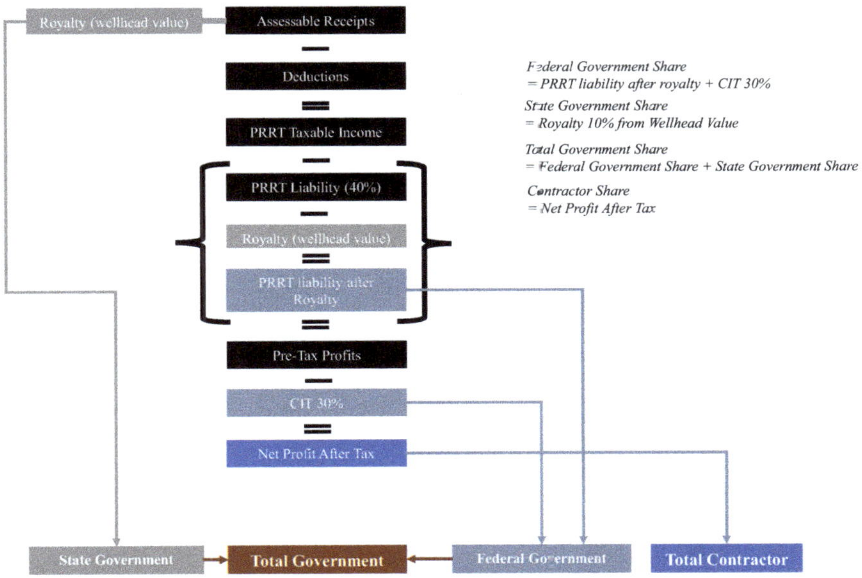

Fig. 6.1 Economic structures of a combined fiscal regime, Australia. (Source: Created by the author (2020 data from the Australian Government Taxation Office official website, 2019; accessible at www.ato.gov.au))

Overall, the whole fiscal system applied to a project located in a state jurisdiction in Australia was similar to the US (in the State of Wyoming in particular). Instead of having a severance tax or ad valorem tax, Australia had the PRRT liable to the federal government (thus making the Australian fiscal regime different from a general royalty/tax regime as applied in the US). What is distinct between these two systems is the taxation base. In the State of Wyoming, the royalty and state taxes are based on the netback value (which is called a wellhead value in the PRRT system). Meanwhile, in Australia, the wellhead value was only based for the state royalty calculation, in which this royalty was deductible against PRRT liability. The PRRT base was the net revenue after all deductions associated with the project, added with other areas' exploration expenditures. Since the royalty payment was deductible against PRRT liability, it was technically based on the net revenue after all deductions.

Table 6.8 Fiscal system simulation in Australia

		State government	Federal government	Contractor's share
Assessable receipts	600	*Current assessable receipts*		
Deductible expenditures for PRRT purposes	(203)	*Total general expenditures, exploration expenditures, total exploration expenditures (other areas)*		
Notional Taxable Receipts (PRRT)	397			
PRRT 40%	(159)			
Royalty 10%	(40)	*40*	*10% from wellhead value assumed to be $400*	
PRRT after royalty	(119)		119	
Net revenue available for CIT	238			
Taxable income	238			
CIT 30%	(71)		71	
Net profit after tax	167			167
Economic profits				
Assessable receipts	600			
Allowable deductions	(203)			
Economic profits	397			
Division of profits				
State government (royalty)	40	10%		
Federal government (PRRT and CIT)	190	48%		
Contractor	167	42%		
Total profits	397			
Government take	58%			
Contractor take	42%			

An additional assumption on this calculation was the allowable deductions for CIT purposes

Source: Simulated by the author (2020—data from the Australian Government Taxation Office official website, 2019; accessible at www.ato.gov.au)

6.5 Fiscal Terms Comparison: Indonesia, the US and Australia

This section will compare the three different fiscal regimes, both qualitatively and quantitatively. The comparative study is conducted by comparing the purpose and impact of the implementation of each fiscal term in each country. This part will focus on the purposes and impacts of the PRRT and the Wyoming fiscal terms.

6.5.1 Qualitative Comparative Study

In contrast to Indonesia, the Australian and the US governments do not have a cost recovery term in their fiscal regime. Australian applied the PRRT (for onshore projects) directly to the resources before the CIT, and the PRRT was treated as an income tax deduction. This PRRT was front-end loaded for the investor, but the investor had the freedom to control and manage their expenditures. There was no control over the expenditures by the government as there was no cost recovery in the system. The government also does not take exploration risk into account.

The fiscal regime in Australia has changed, and in July 2019 an amendment was made on the implementation of PRRT. Before 2012, the PRRT only applied to offshore projects with an exception on the NWS offshore project which applied royalties instead of PRRT. After 2012, the PRRT was only applied to onshore projects and in July 2019, the PRRT has applied again to offshore projects only (after the amendment took effect in July 2019). The PRRT had allowed the onshore project contractor to recoup the exploration expenditures from one group project to another group project in Australia, where it had notional profits (PRRT Taxable Profit). The ring-fence policy only applied to specific expenditures related directly to the project (production project) such as development expenditures and operating expenditures. Both the Australian and the US governments applied a fiscal regime where it contributed directly to the local government (in this case the state governments). It does not work the same in Indonesia since the central government will collect all the possible incomes from oil and gas activities. There is no specific discussion on how the local government (state government) distributes to the local community. The direct contribution paid to the local government was in the type of royalty and taxes that only applied to the specific state. Indonesia applies a 10% local participation for the local government which is expected to create a sense of belonging of the local government to the project. This is not exactly the same as the contribution of the investor to the local government since it requires the local government to participate and in return to receive the 10% share.

The PRRT and R/T systems are relatively easy to administer compared to the PSC applied to CBM in Indonesia. The government provides flexibility to the investors to manage their budget and expenses without neglecting their responsibility for the health, safety, and environment. As mentioned earlier, there are many procedures attached to the PSC system in Indonesia. These procedures are mostly attached to the cost recovery system, meaning that everything has to comply with the procedures, including the technical requirements for CBM activities in Indonesia.

The PRRT system allowed the investor to uplift their carry forward cost at a certain percentage. This was intended to accommodate the loss on the time value of money from the unrecouped costs in that year. This applied to all expenditures, such as development costs, exploration costs; and exploration costs from other group projects in Australia. Indonesia does not apply this type of uplift, but they do provide incentives with an uplift rate which is called investment credit. Technically, they uplift the development expenditures to be recovered, but the additional amount recovered is subject to tax (Table 6.9).

6.5.2 Quantitative Comparative Study

This section looks closely on how exactly the terms impact the governments and contractor's shares. A sensitivity analysis was conducted by changing the revenue and expenditures over time.

The PRRT system was designed to give stability to both the government's and the contractor's shares, but it has been unsuccessful in that aim (for more detail, see discussions on 'Changes in Gross Revenues' and 'Changes in Expenditures' below). Meanwhile, both the R/T and the CBM PSC systems do not guarantee the shares between both of them. Both are regressive; the higher the revenue, the lower is the government share. However, these three fiscal regimes provide confidence to the government since they guarantee the income from royalties (to the state government) and FTP. In the R/T, there is some additional defi-

Table 6.9 Summary of the comparative fiscal regimes applied to CBM in Indonesia, the US, and Australia

Fiscal terms	Indonesian PSC	The US R/T	Australian PRRT	Explanations
Gross revenue/ netback value/ wellhead value and assessable receipts	Gross revenue	Netback price	Wellhead value and assessable receipts	The gross revenue is the FTP-based calculation, which is the value of the total gross production. In the US, the base for royalty calculation is called netback value, while in Australia it was called wellhead value. Both netback value and wellhead value are technically the same, which is the value of the production after all deductions to get the production to the point of sales. However for onshore projects, Australia had assessable receipts which included all type of receipts related to the project and this was the based for the PRRT calculation and finally the CIT base.
Royalty	–	√	√	Indonesia does not have royalty in their PSC, but the FTP which technically calculated from the same based as royalty is calculated (in both R/T and PRRT systems). While in the US and Australia, they apply royalty, which in this particular case (Wyoming State and Queensland State) is paid to the state government. The difference is the calculation basis. The FTP base is the gross production (right at the top), while this royalty is calculated from the wellhead value after all deductions (all cost incurred until the point of sale of the gas). In the US, this based is called the netback value and in Australia it is the same.
(FTP) First Tranche Petroleum	√	–	–	

(continued)

Table 6.9 (continued)

Fiscal terms	Indonesian PSC	The US R/T	Australian PRRT	Explanations
Severance tax/ad valorem tax	–	√	–	These taxes are imposed on a project in the State of Wyoming and are payable to the state government.
PRRT	–	–	√	The Australian fiscal regime does not impose these taxes, but has a 40% PRRT. The difference between Wyoming and Australia is the basis for the taxes. In Wyoming, the basis for calculating the taxes is the value after royalty payment; meanwhile, for Australian onshore projects, the PRRT was calculated based on the value after royalty costs and after expenditures OPEX and CAPEX and exploration costs. The PRRT had become an insignificant revenue source for the Australian government as a result of incentives given to the investors (such as uplift, loss carry forward) before imposing the PRRT. The PRRT for onshore projects was payable to the federal government of Australia. In the US, these taxes are paid to the state government, and PRRT was paid to the federal government.
Cost recovery	√	–	–	Indonesia applies PSC, which is tied to cost recovery mechanisms and requires the investors to comply with specific procedures (applied to conventional gas) which are not necessary for CBM (technically). This mechanism contributes to high cost on CBM project, both from financial and time perspectives.

Uplift rate	√	–	√	The mechanism of the uplift rate in the PSC is different from the mechanism of the uplift rate in PRRT system. The uplift rate in the PSC (which is also called as investment credit) is to incentivise investor by giving an additional development expenditure, and this addition is subject to tax. Meanwhile, in the PRRT system for onshore projects, the uplift rate was a rate given for the investors on the excess of expenditures against the assessable receipts and compounded in subsequent years. Both have different impacts on the economic structures of the project. In the PSC, it will increase the cost recoverable amount in that particular project and will contribute to the government tax income. This uplift rate will shorten the payback period of the investment which will attract investors as having a shorter payback period is one of the investors' interests. In the PRRT, the uplift rate was to compensate the loss of value of money when the investor could not recoup for expenditures in that particular year which led to a lower PRRT liability.
Transferrable exploration expenditures	–	–	√	The Australian government provided an incentive to investors by enabling them to transfer their Australian group project's exploration expenditures to another project that had notional PRRT taxable profit. However, for onshore projects, they applied ring-fence policy between group projects for development and operating costs. Indonesia applies a POD basis ring-fence policy which prevents investors from transferring any costs from a different project to another profitable project (developed).

(*continued*)

Table 6.9 (continued)

Fiscal terms	Indonesian PSC	The US R/T	Australian PRRT	Explanations
Ring-fence policy	√	–	–	The purpose of the ring-fence policy is to protect the government's tax revenue. This policy does not incentivise any investment activities, especially in the upstream industry, which is a high-risk project. For the CIT purposes, the Australian government allowed the investor to consolidate their total income/losses from projects within Australia only.
Tax credit	–	–	√	The Australian government provided an incentive to the investors, a tax credit when they could not cover the closing-down expenditures for onshore projects. This incentive did not apply to any project that had never paid PRRT.
Corporate income tax	√	√	√	All of them impose CIT on their projects.

Source: Created by the author (2020)

nite income from the severance tax and ad valorem tax to the state government. However, this system does not guarantee the federal government to collect the CIT since the contractor may not have enough after paying the state's tax revenue to recoup their expenditures.

1. ***Changes on Gross Revenues***

Gross Revenue VS Government's Shares and Contractor's Shares
This section stipulates how changes in the gas revenue impacts the government's share under PSC, R/T, and PRRT.

Table 6.10 is the assumption table with estimated gas production from a CBM project in Indonesia. However, the numbers have been modified and are slightly different from the actual case to protect the data.

The production profile is 100 bcf for ten years, and the development cost is US$1.8/mcf, and the operation and maintenance (including ASR) costs are US$1.2/mcf. The gas price is assumed to be US$6.15/mbtu or US$6/mcf according to the GR 40/2016.

Table 6.10 shows that there are changes in revenue from year 1 to year 10. The point here is to look at the impact on the government's shares and the contractor's shares when the revenues change, how regressive the system is for the government, and how progressive it is for the contractor under the PSC and R/T systems (Fig. 6.2).

Table 6.10 Ten years gross revenue (changes overtime) and costs profiles (constant) (in million US$)

Year	1	2	3	4	5	6	7	8	9	10
Gross revenue from gas	$60	$65	$70	$75	$85	$70	$65	$60	$55	$60
Transportation costs etc. to point of sales	$8	$8	$8	$8	$8	$8	$8	$8	$8	$8
Other receipts	$15	$15	$15	$15	$15	$15	$15	$15	$15	$15
Deductible expenditures	$22	$22	$22	$22	$22	$22	$22	$22	$22	$22

Source: Estimated by the author (2020—actual data from CBM investor in Indonesia)

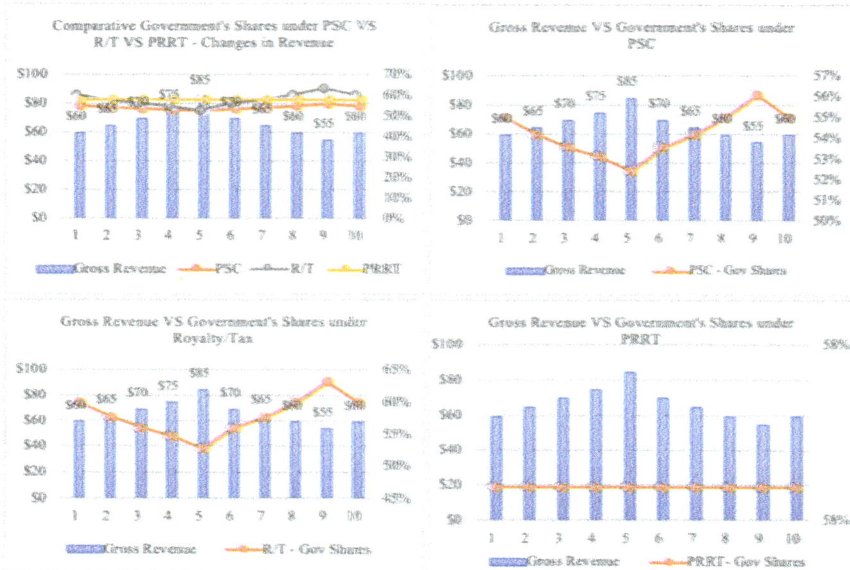

Fig. 6.2 Sensitivity analysis on revenue changes towards government shares using the PSC, R/T, and PRRT. (Source: Created by the author (2020))

According to the figures above, both PSC and royalty/tax systems are very regressive, meaning when the revenue increases, the government share decreases. Unlike the PSC and the R/T systems, government shares under PRRT do not change when the revenue increases. A sliding scale royalty for the royalty/tax system and a sliding scale split for the PSC can be the solutions for regressive systems like the R/T system and PSC.

Unlike the government's shares, the increase in gross revenue gives higher shares for the contractor under the PSC and R/T systems. While in the PRRT system, the government's shares and the contractor's shares are supposed to be constant. Neither changes in the gross revenue (see also Fig. 6.3) and expenditures influenced the contractor's shares under PRRT (see Fig. 6.4).

It can be concluded that the PSC and R/T systems are regressive for the government, as when the gross revenues increase, the government's shares decrease and the contractor's shares increase. As mentioned earlier,

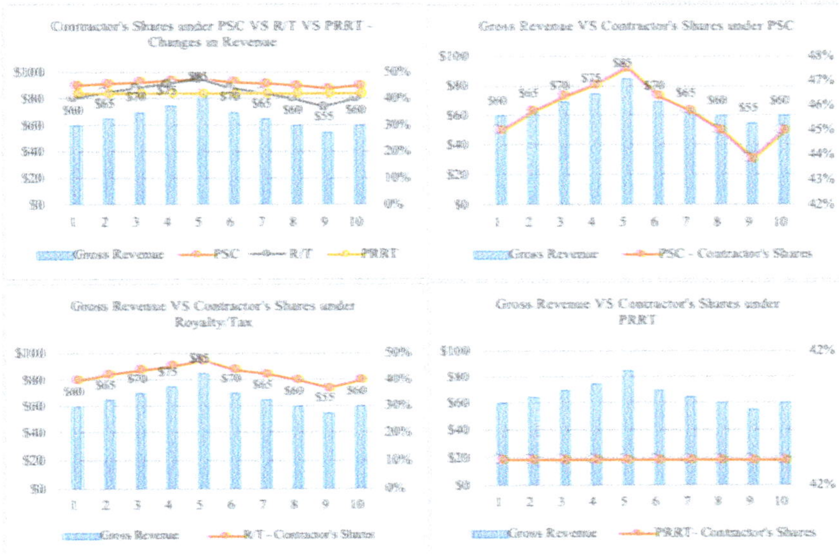

Fig. 6.3 Sensitivity analysis on revenue changes towards contractor's shares using the PSC, R/T, and PRRT. (Source: Created by the author (2020))

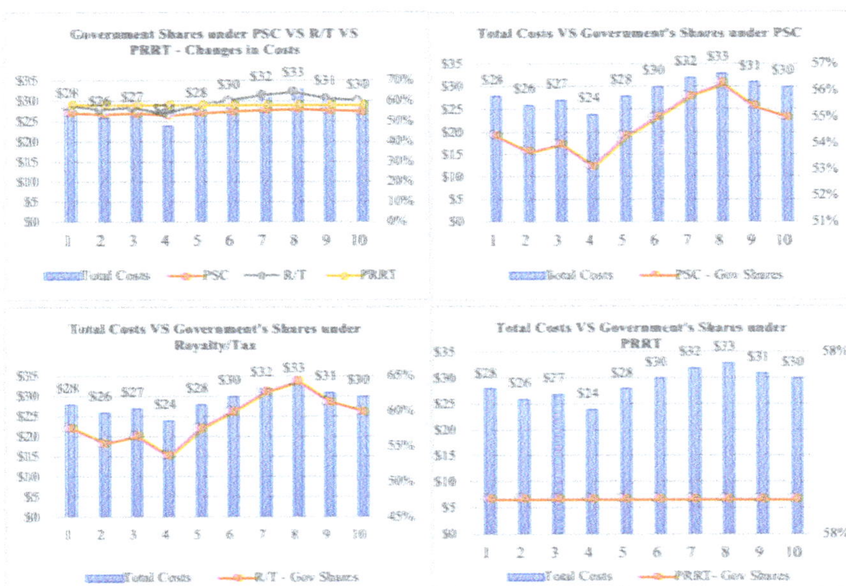

Fig. 6.4 Sensitivity analysis on costs changes towards government shares using the PSC, R/T, and PRRT. (Source: Created by the author (2020))

a sliding scale can be applied to maintain the balance on the shares between the government and contractor.

However, with the same revenue assumptions, the PSC has the lowest contribution to the government shares, and the royalty/tax can give a higher rate compared to the PRRT only when the revenue is low.

The only guaranteed revenue for the government under PSC is the 10% FTP, and the rest depends on the amount of the revenue and cost recoverable items. Meanwhile, in the R/T, the state government is guaranteed to receive royalty and the local taxes (severance tax and ad valorem tax) before the development expenditures and operating expenditures. The federal government is not guaranteed to receive income from the project if the deductions are more than the net revenue after royalty and local taxes. In this case, regardless of how regressive the system is, the R/T is better for the government since the government will receive guaranteed revenue from two different sources which are royalty and the local taxes. While in the PSC, the government is only guaranteed from the FTP.

2. ***Changes on Expenditures***

Total Costs VS Government's Shares and Contractor's Shares

It has been discussed above on how the revenue impacts to the government shares. This section looks at the costs impacts on the government shares. The data assumption is modified data from the actual data with constant revenue and changes on costs over time. The following table is the data assumption used to do this sensitivity analysis (Table 6.11).

Figure 6.4 shows that PRRT is supposed to provide constant shares to the government regardless of changes in costs. The changes in revenue or costs are not supposed to impact the government's shares in the PRRT system. Under the PSC and R/T systems, the changes in costs do impact on the government share.

Unlike the gross revenue, the increase in costs creates a better share for the government under the PSC and the royalty/tax systems. This does make sense since the government's share is regressive towards the net

Table 6.11 Ten years gross revenue (constant) and costs profiles (changes overtime) (in million US$)

Year	1	2	3	4	5	6	7	8	9	10
Gross revenue from gas	$60	$60	$60	$6C	$60	$60	$60	$60	$60	$60
Transportation costs etc. to point of sales	$8	$8	$8	$8	$8	$8	$8	$8	$8	$8
Other receipts	$15	$15	$15	$15	$15	$15	$15	$15	$15	$15
Deductible expenditures	$20	$18	$19	$16	$20	$22	$24	$25	$23	$22

Source: Estimated by the author (2020—actual data from CBM investor in Indonesia)

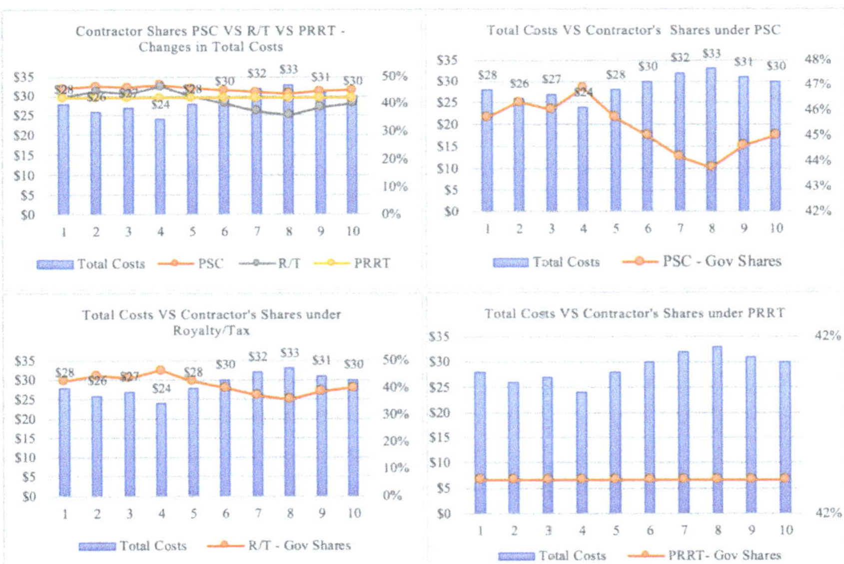

Fig. 6.5 Sensitivity analysis on costs changes towards contractor's shares using the PSC, R/T, and PRRT. (Source: Created by the author (2020))

income. The higher the costs with a constant revenue, the lower is the net income which then provides the government with a better share. After looking at the government's share perspective, the following section will discuss the contractor's share perspective.

Figure 6.5 demonstrates how the contractor's shares increases or decreases when the cost changes over time in that project. Under the PSC and R/T systems, an increase in gross revenues brings an increase in the contractor's shares; on the contrary, an increase on expenditures brings a decrease in the contractor's shares.

The table shares both the government's and the contractor's sides under PRRT as constant. This is because the split was after all deduction. As we can see from the earlier tables—Tables 6.5 and 6.6—on the *Economic Structures of the Gas in State Land* (Wyoming R/T), some payments to the government are before the deductions (royalty and the local tax), and similar to the PSC, the FTP is paid before the deduction (taken from the gross revenue). In conclusion, if the split between the government and contractor was based on the net revenue, it would provide stability on the shares (see Fig. 6.1). Meanwhile, when some payment or split to the government is based on the gross revenue (before all deductions), it would create unstable shares when there are changes in costs or revenue.

The analysis above will help to draw a conclusion on what fiscal terms to be adapted, which will be discussed in the next section.

6.6 Feasible Fiscal Terms to Be Adapted

This section discusses what can be adapted from the two countries to the Indonesian fiscal regime. It also looks at the literature on developing fiscal regimes. The possible outcome of the proposed fiscal terms is to accommodate issues found in the previous chapters.

Table 6.12 provides what can be adapted and what issues are being accommodated by implementing the terms into the Indonesian PSC. The previous chapters of this book discussed the issues that need to be resolved to allow the CBM industry to grow in Indonesia:

1. Government is being ignorant towards the industry.
2. A fiscal regime that requires the contractor to comply with complex regulation and procedures which do not accommodate the technical differences between conventional gas and CBM gas.
3. Technical issues related to the nature of the CBM geology in Indonesia. This can be resolved by having a fiscal regime that takes the technical requirements into consideration.

Table 6.12 Feasible terms to be adapted for Indonesian CBM PSC

Fiscal regimes	Terms to be adapted	Issues being accommodated	Remark	Implementation
Royalty/tax	Royalty system	Local government and local community's income distribution	The PSC does not apply royalty but, the FTP. The FTP is taken from the top (gross revenue). In the US, the royalty for resources located in a state belongs to the state government and not to be shared with the federal government, and royalty for resources located in the federal land should be shared between the federal government and the local government.	Indonesia may adopt the system but instead of having the FTP payment to the central government, it should be shared between the local government and central government.
Royalty/tax	Severance tax and ad valorem tax	Local government income and local community income distributions	The severance tax and ad valorem tax are both imposed before the deductions. This will guarantee that the local government will receive the contribution for having the energy development.	Indonesia may not implement the severance tax and ad valorem tax, but with the same mechanism, Indonesia can impose taxes related to the local community development before the cost recovery.

(*continued*)

Table 6.12 (continued)

Fiscal regimes	Terms to be adapted	Issues being accommodated	Remark	Implementation
Royalty/tax and PRRT	Flexibility (technical requirements)	Cost recovery issues and accommodating the technical issues that were not being accommodated in the current PSC	One of the main issues in the PSC is the cost recovery procedure. It creates complexity and costs investors time and money. This will protect the investors from losing time and money due to the complexity and the strict requirements (which sometimes is not required for CBM project)	The US does not have the cost recovery mechanism in their fiscal arrangements. Hence, it gives flexibility for investors on how they will operate their activities provided they comply with all other regulation such as environment and HSE regulation. Having this flexibility will shorten the time they require to start the operation. Indonesia applies PSC for CBM, meaning that the investors should consider obtaining approval from the GoI to do any activities. Cost recovery procedures should not prevent the investor from doing their activities. Having procedures for the cost recovery purposes are important, but it should be more flexible according to the technical assessment. However, the HSE should always be put as a priority in the approval process. GoI shall standardise what important matters and not applying a strict requirement to the technical or technology being used to run the project. Simplified procedures for approval should be developed in the system.

PRRT	Ring-fence policy[a]	CBM technical issues	Protect investors cash flow and country development	Indonesia could implement similar ring-fence policy as applied in Australia. A CBM project requires a hundred wells to drill. Therefore, ring-fence may not be applied for exploration expenditures but may be applied for development expenditures and production cost (OPEX). Instead of applying a ring-fence in each field, Indonesia could apply the ring-fence policy only for development and expenditures and allows exploration expenditures to be offset to a profitable project.
PRRT	Uplift rate[b]	Financial recoverability for investors and local community development	This will encourage the investors to do more investment, since they are aware that their expenditures are being uplifted to mitigate the loss due to the time value of money and to cover other costs that are not considered as qualified expenditures.	Indonesia can have a similar mechanism by applying the uplift rate to specific expenditures. Indonesia has applied the uplift rate (called investment credit), but this uplift rate is fixed and subject to tax. The uplift rate in Australia was applied to carry forward losses for project cost, but for onshore projects, the rate proved to be too generous. Indonesia could adopt this uplift rate mechanism for any carried forward expenditures on CSR activities prior to production. This would encourage investors to develop the community as they would receive this incentive from the GoI.

(continued)

Table 6.12 (continued)

Fiscal regimes	Terms to be adapted	Issues being accommodated	Remark	Implementation
Others	Sliding scale split[c]	Balancing fiscal system (progressive vs regressive)	Protect both government's and investors' cash flow	It is important in Indonesia to ensure that the government does not lose financially yet to also keep the attractiveness of the system for the investors. Therefore, a sliding scale system can help to prevent the government being seen as only giving the benefits to the investors.

Source: Created and analysed by the author (2020)

[a]However, following the Treasury Amendment on PRRT, 1 July 2019, Australia has now removed the ring-fence policy

[b]Australian Government has lowered the uplift rate (Treasury Amendment on PRRT, 1 July 2019) as they were considered too generous. What the author was trying to recommend is to adopt the mechanism, meanwhile, for the rate will require a further research and analysis

[c]The author is aware that sliding scale split can create complexity on its implementation and may not raise tax revenue for the GoI. It should be noted, that this research suggested that the GoI should return to their energy paradigm that "the energy resources are no longer as export commodities only, but as the engine of national growth" (Article 6, GR 79/2014). By having a more attractive fiscal arrangement, more investors would invest and develop the local community, and the GoI eventually is involved indirectly in economic growth

4. Local community issues which are very much related to community development and the expectations of the local people.
5. The financial issues faced by the contractor.

All of the issues above can be resolved by having a good fiscal regime that fits into the key stakeholders' perspectives (not only government and investors but also the local community). Indonesian can learn from the mechanisms and apply it contextually. Table 6.12 provides the terms and mechanisms from the different the R/T and PRRT systems that can be adopted and what issues are being accommodated. Unlike the proposed recommendation stated by the NEC, this recommendation is limited and more specific to the fiscal terms.

6.7 Discussion on Feasible Terms to Be Adapted

As mentioned earlier in Table 6.12, discussions mainly focus on the contribution to economic development through local communities. First, the royalty and severance tax mechanisms are expected to contribute directly to local government revenues. Second, flexibility is needed to accommodate the technicalities of CBM. Having flexibility will address the cost recovery issues and technical challenges facing CBM. Third, the uplift rate mechanism will be adopted from Australia and applied it to the CSR expenditures instead of the exploration costs, as long as the uplift rate is not too generous. Fourth, the sliding scale will balance the government's and investor's cash flow.

6.7.1 Contribution to the Local Community

The royalties and local tax (severance tax and ad valorem tax in the R/T system) being proposed in this research are for direct contribution to the local community. The contribution to local community development has a positive relationship with the company's profitability (Frynas 2009). It will drive the local government to support the CBM projects in their area since their productivity will allow them to receive a direct contribution from the investors.

(i) ***Royalty***

The royalty system in the State of Wyoming is applied based on where the oil/gas is located, either federal land or state land. As previously explained, if the oil/gas is located on federal lands, the royalty payment is paid to the federal government, but the state government will also get a contribution from the royalty payment. Indonesia does not have a royalty rate in the PSC, but Indonesia has FTP, which technically is a royalty. In Indonesia, all land and water in the area of Republic of Indonesia belong to the central government. However, the FTP payment is always made to the central government. This research has identified issues in the CBM industry which are directly related to the local communities. Kristiansen and Pratikno (2006) stated that one of the local government revenue sources is from the contribution of natural resources. Therefore, this research proposes an idea of splitting the FTP for the central government and the local government

(ii) ***Severance Tax and Ad Valorem Tax***

Local government should have an active role in managing local taxes to achieve social and economic objectives (Aničić et al. 2016). The severance tax and ad valorem tax are usually known as local taxes (Johnston 2018). These taxes are taken before all the deductions but after royalty. This royalty is to protect the local government revenue generated in the energy sector (in this case is resource extraction). Indonesia does not apply these local taxes in the PSC for CBM. Therefore, this research proposes to implement the local taxes as an insignificant percentage to provide a direct contribution to the local community and help the local government to achieve their social and economic aims.

6.7.2 Accommodating Cost Recovery Issues and Technical Issues

PSC for CBM in Indonesia has created problems for investors to develop their CBM projects. This is mainly related to the cost recovery mechanism as it is very much linked to conventional hydrocarbon procedures which are not applicable for CBM. This research identified two key changes in the PSC terms to deal with the cost recovery and technical issues which are cost recovery flexibility and the ring-fence policy (from the PRRT, Australia). In the end, by encouraging the investors, the country will benefit from local development from the community development activities by the investors.

(i) ***Cost Recovery Flexibility***

This research proposes flexibility in terms of the cost recovery mechanism following the technical difficulties in the CBM project that do not exist for conventional hydrocarbon projects (e.g. rig specification for a CBM well does not have to follow the rig specifications for conventional hydrocarbons since the well being drilled is shallower and thus does not necessarily require hiring a huge rig which will cost much more for the CBM project). This flexibility will improve the attractiveness of the PSC from the investors' point of view (Wajong 2018; Setyatmoko 2018).

(ii) ***Ring-Fence Policy***

The ring-fence policy being proposed in this research is to protect the government's and investors' cash flow. Indonesia applied a ring-fence policy in its PSC system, which does not allow exploration expenditures in dry hole fields to the production field. This research proposes a similar implementation as there was for onshore projects' OPEX and CAPEX costs in the PRRT system in Australia. However, the Australian design failed as the uplift rates were too generous. A CBM project requires more than 100 wells to drill in the exploration phase. Having this ring-fence policy implemented only for development expenditures and production costs could encourage the investors to do more exploration and contribute to the countries reserves/resources data (whether or not they are successful) and development in the energy sector (if they are successful).

6.7.3 Balancing the Investors' and Local Community's Interests

This research identifies the uniqueness of the PRRT system in Australia that is related to the uplift rate system. The uplift rate system applied to the PRRT system was not the same as the uplift rate applied to the Indonesian PSC. The Indonesian PSC applied an uplift rate only for development expenditures, in which the investors can add their recoverable cost item as much as the uplift rate (but this uplift rate is taxable). In contrast, in Australia the uplift rate was applied to onshore projects' carry forward expenditures, which accommodated the loss on the time value of money. This research proposes the adaptation of the uplift rate mechanism of the PRRT, not to be applied in the carry-forward expenditures, but only to all expenditures related to social costs and CSR (community development). This would encourage investors to spend on the local community development. While the local community can enjoy the benefits of having a CBM project in their area, the investor can also sustain their project and improve their financial performance (Freilich and Popowitz 2012; Streimikiene et al. 2009; Orlitzky et al. 2003; Busch and Friede 2018).

6.7.4 Balancing the Government and Investor's Cash Flow

A sliding scale split is proposed in this research as it is known as a mechanism to balance both government and investor cash flow.[3] In principle, the government will have a better split when the production

[3] The author is aware that sliding scale split can create complexity on its implementation and may not raise tax revenue for the GoI. It should be noted that this research suggested that the GoI should return to their energy paradigm that "the energy resources are no longer as export commodities only, but as the engine of national growth" (Article 6, GR 79/2014). By having a more attrac-

increases, meaning that at the beginning of the production stage the investors will have a better split (in percentage) than the government but gradually this will decrease following increases in production. This will attract the investors, as they will have a better offer when their production is low, and when the production high, they will have less percentage but not necessarily experience a lower cash flow.

Finally, it is significant for the government to understand the impact of each of the terms that they apply to their fiscal regimes. Table 6.12 is developed contextually by looking at the two other fiscal regimes as related to CBM.

References

ALL Consulting, 2004. *Coal Bed Methane Primer*. New Source of Natural Gas – Environmental Implications. Montana: U.S. Department of Energy, National Petroleum Technology Office.

Aničić, J., Jelić, M. and Đurović, J., 2016. Local Tax Policy in the Function of Development of Municipalities in Serbia. *Procedia – Social and Behavioral Sciences*, 221, pp. 262–269.

Busch, T. and Friede, G., 2018. The Robustness of the Corporate Social and Financial Performance Relation: A Second-Order Meta-Analysis. *Corporate Social Responsibility and Environmental Management*, 25(4), pp. 583–608.

Daniel Johnston & Co., Inc, 2018. Wyoming: Legal and Fiscal Frameworks: Best Practice. Analysis and Commentary: Fiscal System Study. [online] Available at: http://www.wyia.org/2018/12/03/oil-gas-tax-study-legal-fiscal-frameworks-best-practices/[Accessed 1 May 2019].

Freilich, R. and Popowitz, N., 2012. Oil and Gas Fracking: State and Federal Regulation Does Not Preempt Needed Local Government Regulation. *The Urban Law*, 44(3), pp. 534–575.

Frynas, J., 2009. Corporate social responsibility in the oil and gas sector. *The Journal of World Energy Law & Business*, 2(3), pp. 178–195.

Jackson, H., n.d. *Drilling for Tax Credits: Runaway Coalbed Methane Development Doesn't Need a Tax Break*. [ebook] Public Citizen. Available at: https://www.citizen.org/sites/default/files/coalbedfactsheet.pdf [Accessed 24 April 2019].

tive fiscal arrangement, more investors would invest and develop the local community, and the GoI eventually is involved indirectly in economic growth.

Johnston. D., 2018. *Wyoming: Legal and Fiscal Frameworks: Best Practice. Analysis and Commentary: Fiscal System Study*. [online] Available at: <http://www.wyia.org/2018/12/03/oil-gas-tax-study-legal-fiscal-frameworks-best-practices/> [Accessed 1 May 2019].

Johnston, D., 2019. *Severance & Property Tax & Ad Valorem Tax*. [email].

Kristiansen, S. and Pratikno, 2006. Decentralising education in Indonesia. *International Journal of Educational Development*, 26(5), pp. 513–531.

Orlitzky, M., Schmidt, F. and Rynes, S., 2003. Corporate Social and Financial Performance: A Meta-Analysis. *Organization Studies*, 24(3), pp. 403–441.

PricewaterhouseCoopers, 2015. *Oil and Gas in Indonesia*. Investment and Taxation Guide, May 2014 — 6th edition. Jakarta: PricewaterhouseCoopers.

Setyatmoko, U., 2018. *A Focus Group Discussion on CBM Industry in Indonesia*.

Streimikiene, D., Simanaiciene, Z. and Kovaliov, R., 2009. Corporate Social Responsibility for Implementation of Sustainable Energy Development in Baltic States. *Renewable and Sustainable Energy Reviews*, 13, pp. 813–824.

The Wyoming Legislature, 2019. [online] Wyoleg.gov. Available at: https://www.wyoleg.gov/InterimCommittee/2015/SMTPresentationAG-DoR.pdf [Accessed 11 July 2019].

Tordo, S., 2007. *Fiscal Systems for Hydrocarbon, Design Issues. 1st ed. [ebook] Washington: The World Bank*. Available at: https://siteresources.worldbank.org/INTOGMC/Resources/fiscal_systems_for_hydrocarbons.pdf [Accessed 24 April 2016].

Treasury Laws Amendment (2019 Petroleum Resource Rent Tax Reforms No. 1) Act 2019. Royal Assented to 5 April 2019.

Wajong, M., 2018. *A Focus Group Discussion on CBM Industry in Indonesia*.

7

The Future of CBM in Indonesia and International Reflections

Abstract This chapter concludes the book, exploring in particular the future outlook of the energy transition in Indonesia and what role CBM can play in energy transition and energy sustainability. As such, it will examine the dos and dont's for the government in developing the clean energy project as a whole and in supporting CBM development in the country. Finally, this chapter provides a number of recommendations to the GoI to achieve future energy transition goals out towards 2050.

Keywords CBM future • Energy development • Fiscal issues • First Tranche Petroleum • Environmental tax • Sliding scale split • Ring-fence policy

7.1 Introduction

This chapter concludes the book by exploring the future outlook of the energy transition in Indonesia and what role CBM can play in energy transition and energy sustainability. It will examine the dos and don'ts for the GoI in developing the clean energy project as a whole and in

T. B. Sumarno, *Just Energy Transitions and Coal Bed Methane*, Energy, Climate and the Environment, https://doi.org/10.1007/978-3-030-85490-4_7

supporting CBM development in the country. Finally, this chapter provides recommendations for the government in achieving the energy transition goal of 2050.

7.2 CBM in Indonesia's Future

By comparing the CBM cost structures and fiscal regimes of three different countries via a comparative analysis, this book has provided new knowledge on how technical and geological conditions can influence the costs of CBM projects. Learning lessons from the current fiscal regimes of royalty/tax system in Wyoming (US) and the PRRT system in Queensland (Australia) have revealed the terms to be adopted to address the current challenges which are interconnected. This is part of the international context that can be adopted to the national context, that is, Indonesia.

Previous literature only discussed the PSC in general; this book has taken it a step further by analysing it quantitatively to identify the CBM challenges. Further, the uniqueness of this research is that it involves most of the stakeholders of the CBM development and obtains data from these stakeholders directly to discover the real challenges in the industry. Data and information were obtained from the internal (government, investors, local communities, industry association) and external (global consultancy) perspectives. In addition, this research also utilised the investors' perspectives. It was also critical to underline the importance of the local community role in project sustainability and the need to directly engage with local communities to obtain real-world data on the impacts that they have felt from the CBM development and their expectation towards CBM development. The results showed that each area has different priorities regarding their expectations and impacts which can help the contractor in developing a strategy for project sustainability.

Overall, then, this book has comprehensively discussed the GoI's energy goals and how CBM can help to achieve the set goals, by identifying the CBM challenges from different perspectives, and defining what terms of the fiscal regime can be modified to address the challenges. Each of these processes has contributed to new knowledge for academia, government, and industry.

7.3 The Core Energy Issues for Energy Development

The first core issue in terms of energy development for CBM is about the influence of geological conditions to the current CBM fiscal regime (PSC) in Indonesia and how the current PSC has failed to support the growth of the CBM industry in Indonesia. Chapter 3 identified the CBM cost structures in Indonesia which have been very much influenced by the geological conditions in Indonesia. The cost of dewatering management is huge in that it is considered as capital expenditure instead of operating expenditure. This is due to the climate and more likely due to the water content in the coal. It takes longer to pump out the water, and it costs more to process the water before it can be released to the environment. CBM gas is not stored in the same depth as the conventional gas yet the PSC requires investors to apply the same technical requirements to both of them. Therefore, CBM gas production in one well fluctuates substantially, and it requires hundreds of wells to be drilled to make the project commercially viable. The latter needs to be taken into consideration when applying the fiscal regime to the CBM, since it is technically and geologically different from conventional gas.

Qualitatively, it has been found that the PSC failed to appropriately address the CBM geology aspect, and thus failed to support the growth of CBM as an industry. The fiscal terms and procedures created complications resulting in higher costs. It was also time consuming. Put bluntly, CBM gas is not economically viable if it has to comply with the PSC. The Gross PSC Sliding Scale (SS) considered by some CBM investors attempts to create more cost efficiencies for the project. However, the analysis shows that this PSC failed to take low gas prices into account. In other words, the Gross PSC SS only works well if the gas price is high.

The second major issue in terms of energy development concerned the issues in CBM development in Indonesia. The investors found a fiscal regime to be attractive if:

1. It allows the investors to create a minimum return on their investment. This is a common standard for any fiscal regime. However, it is not only profitability that the investors find attractive.
2. A fiscal regime should not be complicated to comply with. Unlike conventional gas, a CBM project requires extra activities to extract the gas. Therefore, a simplified fiscal regime with simplified rules and procedures will be more attractive for investors.
3. The investors find it attractive when a fiscal regime allows the local community to be part of the project or directly benefit from the project. This type of fiscal regime will support the investors indirectly by retaining its project sustainability in the local area.
4. A fiscal regime that puts forward the economic development of the area rather than revenue generation of the central government will be more attractive for the investors. The investors can contribute directly to the local community by having a community development strategy. Government shall consider the non-financial impacts of the energy development in a local area (especially one that is isolated).

From a government perspective, it is difficult to accept the type of fiscal regime that does not contribute to revenue generation for the central government. In the current context, it is not possible to have that type of fiscal regime. Therefore, the GoI must be reminded of the energy paradigm that they have set in the Government Regulation No. 79/2014, Article 6. Indonesia is a big nation with many islands which are not equally developed. The government can take the advantage of having investors developing isolated areas. The fiscal regime shall still bring revenue for the government, but it shall not be the main priority of the government. Chapter 6 identifies what terms are politically acceptable for the government which also address the issues. It can be modifications in the PSC or in the associated rules and regulations. These are detailed below.

(1) *Production-Sharing Contract Modification*

The characteristic of the resources are not perfectly identified at the early stage of the project; therefore, the GoI needs to be aware and allow amendments on the contract in accordance with the fact that technical and resources characteristics are found after some period of time. A valid and scientific justification is required for enabling the contractor to amend the terms of the PSC, for example incentives on uplift rate (see below for explanation).

The following points are specific terms and modifications that need to be taken into account:

1. **A shared FTP between the local government and the central government.** Both levels of government will benefit directly from CBM gas production. This will financially help the local government to develop their community. The national government should have this agreed in the contract (PSC) before the investor commences their activities.
2. **An uplift rate for community development expenditures** can encourage the investors to spend more time developing the community, and the government is in charge to monitor these activities to ensure that the investors do the required and proper activities in developing the relevant local communities. This strategy will be more efficient and effective for the government since the government does not spend financially to develop the areas. This will also bring a sustainability strategy to the CBM project in that area since the local community gets direct benefits for having CBM development in Indonesia. This uplift rate has general purposes, but in fact, this can also be due to the unqualified costs that occur during the project life time.
3. The GoI can implement an **environmental tax or local community tax** that is payable to the local government. This tax is to compensate for the negative impacts of the CBM activities to the local community.
4. The GoI can also apply a **sliding scale split** to balance the shares between the government and investors due to fluctuated production (price volatility is not considered in this case as the price is set in the contract and assumed to be a fix price in the contract).

5. GoI can apply the **ring-fence policy for the project based on the POD field**. The GoI can adopt the system from Australia in that it applies the ring-fence policy for the development project, but **not for the exploration expenditures**. This will encourage the investors to conduct more explorations and increase the chance of having gas produced to supply the electricity in the local area.

(2) ***Rules and Regulation***

Rules and procedures for cost recovery are important but simplified standardised procedures will be better for the industry. The GoI can set limitations on each procedure instead of having strict procedures to comply with; and HSE issues are not to be ignored under any circumstances. The GoI should understand better the technical and resource characteristic to enable them to create a distinct and applicable PTK (Working Procedures and Guidelines) for CBM project.

The GoI should again focus on the multiplier impact rather than just temporary revenue, and, in this case, the researcher does not propose any changes on the paradigm being set earlier by the GoI, but reminds the GoI that having investors doing business in the country should bring benefits for the people directly and/or indirectly (by tax).

Finally, the bigger question is what the optimal fiscal regime is for CBM industry in Indonesia for the next 30 years. This research does not produce a new fiscal regime applied to the CBM industry in Indonesia, but it has produced contributions to the idea of modifying the current CBM PSC. A key result is that the optimal fiscal regime for CBM industry (or any resources) does not exist. A fiscal regime can be stated as an optimal fiscal regime for CBM (or any specific resources) when it fits the context and conditions of the resources. A fiscal regime applied to CBM project in Area A, for example, may be an optimal fiscal regime for Area A but may not be an optimal fiscal regime as applied to CBM project in other area/s. There is not one fit for all fiscal regimes. Therefore, prior to negotiation for the fiscal terms, both government and investors need to understand the conditions of the resources as well as the local community situations and conditions.

7.4 Where Next for CBM?

There are other countries involved in CBM projects with different production profiles and cost profiles and modified fiscal regimes. China could represent one such country of interest. However, more data would be needed, as well as access to CBM investors and other stakeholders. More analysis on energy justice and the energy trilemma is needed to assess Indonesia's economic growth through the lens of climate change, environmental degradation, and so on, by including different topics in more detail such as energy/environmental sustainability, energy affordability, and energy security.

Research on how governments (including Indonesia) set the price for the CBM gas to make the project viable will also be another interesting area. In Indonesia, this would be limited to the CBM gas sold to PLN (the Indonesian National Power Plant), which is mostly a coal-based power company. The price required for CBM gas can be much more expensive than coal, but the gas will be used to supply electricity which is the same final product of coal. This price (which is fundamentally a cost for PLN to produce electricity) will blend with the price of the coal, and it is the final price that customers have to pay that matters. This particular research could then touch on issues around energy access for consumers, that is, energy justice.

The domino effect (multiplier effect) of the implementation of the modified fiscal regimes for energy development (CBM or other resources) is also a potential area of interest. This effect can be seen locally (local community) and then brought to a national level. A multiplier effect can be identified in this research which hypothetically can influence government perspectives and bring back the national energy paradigm into policy. Hypothetically, such a fiscal regime would not necessarily contribute to government revenues but instead boost societal development more than governments can do by themselves.

7.5 Final Reflections

Electricity demand in Indonesia is increasing and will only continue to do so. Regardless of having 453TCF CBM gas in the ground, coal has been and remains the major source for electricity generation in Indonesia. CBM projects were initially started in 2008 with initial studies in 2001. However, these projects remain stalled after ten years and are now declining.

The current global trend is moving towards cleaner energy sources over time. In Indonesia, renewable energy alone would not likely be able to supply 270 million people across the thousands of islands that form the country. As a country heavily dependent on fossil fuel–, transitioning to cleaner energy sources could be a challenge for Indonesia. However, CBM could play a role in accelerating the energy transition in Indonesia as it is known as a cleaner energy source (almost like natural gas) amongst fossil energy sources.

Identifying issues that hamper CBM growth is fundamental to this. These issues are interrelated; therefore, solving one issue often fails to solve the rest of the issues, a finding that has also been found for renewable energy technologies (Wood 2013, 2016). Issues need to be addressed in parallel and in a strategic manner. Actions or decisions being made should be supportive of each other. The results here are based on a mixed methods approach involving interviews, focus group discussion, local community survey, and literature reviews. Different perspectives on the CBM industry in Indonesia were obtained from the full range of stakeholders such as government, industry, consultancies, industry associations, and the local communities.

The countries' case studies (Indonesia, the US, and Australia) in this research provide lessons on what can be learnt from each other's experiences and also what terms can be adopted to support the CBM growth in Indonesia. The modifications mentioned in this chapter are contextual for current situations in the country. These are unique to CBM resources in Indonesia in those four different basins. These may or may not be applicable to other countries with CBM resources, as they may have different resources (CBM), characteristics and communities' cultures, and level of development in the community.

This research ends at a point where political changes are happening in Indonesia, and therefore this study is timely. Indonesia has faced many challenges in the energy industry, and will continue to do so given the energy demand from a quickly growing population; hence, this study provides insights for government to develop energy strategy or energy policy. Indonesia is a resource-rich country, and yet to have renewable energy sources developed widely in the country. Indonesia is still looking for more energy in the future to fulfil the domestic needs; therefore, either renewable energy or CBM gas can potentially act as a substitute for the main energy sources. This is why this study is important to support energy development in Indonesia while still developing renewable energy as new resources.

It is interesting to see how a country manages their resources, and they always have space to improve the management in all sectors related to energy and achieve an overall just transition to a low-carbon economy. This research provides proof, which is that for each energy resource and each field there are technical differences that any government shall need to assess before imposing rules, procedures, and taxes. The result may not be suitable for all places, but the role of taxation and justice can be taken into consideration by any country, as well as for any energy resource when developing it.

References

Wood, G. 2013. Connecting the dots: A systemic approach to evaluating potential constraints to renewable electricity technology deployment to 2020 and beyond in the United Kingdom' (PhD thesis, University of Dundee)

Geoffrey Wood, G. 2016. 'A systemic approach to renewable electricity technology deployment: The 'missing link' in optimising policy delivery in the UK?' in Raphael J. Heffron and Gavin F. M. Little (eds), *Delivering Energy Law and Policy in the EU and the US: A reader* (EUP 2016).

References

National Exploration Committee. 2015. Jakarta.

Acim.nidec.com. 2020. *Yangquan Coalbed Methane Power Plant*. [online] Available at: https://acim.nidec.com/generators/leroy-somer/references/yangquan-cbm-power-plant [Accessed 30 July 2020].

adpm.or.id. 2018. *Asosiasi Daerah Penghasil Migas*. [online] Available at: https://adpm.or.id/profil-asosiasi-daerah-penghasil-migas-adpm/ [Accessed 11 September 2018].

Advanced Resources International, Inc., 2002. *Powder River Basin Coalbed Methane Development and Produced Water Management Study*. Tulsa: U.S. Department of Energy.

Advanced Resources International, Inc., 2006. *The Economics of Powder River Basin Coalbed Methane Development*. Prepared for: U.S. Department of Energy. [online] Advanced Resources International, Inc. Available at: http://www.adv-res-hou.com/pdf/The%20Economics%20of%20Powder%20River%20Basin%20Coalbed%20Methane%20Development.pdf [Accessed 27 September 2017].

Al-Jubori, A., Johnston, S., Boyer, C., Lambert, S., Butos, O., Pashin, J. and Wray, A., 2009. *Coalbed Methane: Clean Energy for the World*. Schlumberger.

ALL Consulting, 2004. *Coal Bed Methane Primer*. New Source of Natural Gas – Environmental Implications. Montana: U.S. Department of Energy, National Petroleum Technology Office.

T. B. Sumarno, *Just Energy Transitions and Coal Bed Methane*, Energy, Climate and the Environment, https://doi.org/10.1007/978-3-030-85490-4

Allen & Overy, 2013. *Guide to Extractive Industries Documents – Oil & Gas.* World Bank Institute Governance for Extractive Industries Programme. [online] Allen & Overy. Available at: http://www.allenovery.com/SiteCollectionDocuments/geiprogram.pdf [Accessed 25 November 2016].

Anderson ZurMuechlen and Co., 2001. *Coalbed Methane Development in the Powder River Basin of Montana: Economic and Social Impacts of Proposed Development.* [online] Montana: The State of Montana Government. Available at: https://deq.mt.gov/Portals/112/Energy/CoalbedMethane/Documents/cbm2.pdf [Accessed 19 March 2020].

Anderson, J., Simpson, M., Basinki, P., Beaton, A., Boyer, C., Bulat, D., Ray, S., Reinheimer, D., Schlachter, G., Colson, L., Olsen, T., John, Z., Khan, R., Low, N., Ryan, B. and Schoderbek, D., 2003. *Producing Natural Gas from Coal.* Autumn 2003. USA: Schlumberger, p. 31.

Aničić, J., Jelić, M. and Đurović, J., 2016. Local Tax Policy in the Function of Development of Municipalities in Serbia. *Procedia – Social and Behavioral Sciences,* 221, pp. 262–269.

ANON, 2015. *Project Evaluation for CBM in Indonesia.* [Microsoft Excel] Jakarta.

Apriyani, N., Suharmono, Momen, M., Djaelani, S., Arifin, S., Satria, A., Murtani and Seto, A., 2013. Integrated Cleat Analysis and Coal Quality on CBM Exploration in Sangatta II Field, Kutai Basin, East Kalimantan. In: *Indonesian Petroleum Association 37th Annual Convention.* Jakarta: Indonesian Petroleum Association.

ASEAN Centre for Energy, 2015. *ASEAN Plan of Actions for Energy Cooperation.* Phase I. [online] Jakarta: ASEAN Centre for Energy, p. 3. Available at: https://policy.asiapacificenergy.org/sites/default/files/HighRes-APAEC%20 2016-2025.pdf [Accessed 9 November 2020].

Asian Development Bank, 2020. *Renewable Energy Tariff and Incentives in Indonesia.* Review and Recommendation. [online] Manila: Asian Development Bank. Available at: https://www.adb.org/sites/default/files/publication/635886/renewable-energy-tariffs-incentives-indonesia.pdf [Accessed 11 November 2020].

Ato.gov.au. 2019. *How to Work Out PRRT.* [online] Available at: https://www.ato.gov.au/Business/Petroleum-resource-rent-tax/In-detail/What-you-need-to-know/Work-out-PRRT/How-to-work-out-PRRT/ [Accessed 8 May 2019].

Australian Government Department of Industry, Innovation and Science, 2015. *Review of The Socioeconomic Impacts of Coal Seam Gas in Queensland.* Canberra: The Commonwealth of Australia.

Australian Government Geoscience, 2015. *Offshore Petroleum Exploration Acreage Release*. Canberra: Australian Government Geoscience, Department of Industry and Science.

Australian Government, 2012. *Australia Gas Resource Assessment*. Commonwealth Australia.

Australian Government, 2016. *Petroleum Resource Rent Tax Review*. Commonwealth of Australia.

Australia's Future Tax System Review, 2010. *Australia's Future Tax System Review*.

Bachriadi, D. and Suryana, E., 2016. *Land grabbing and speculation for energy business: a case study of ExxonMobil in East Java, Indonesia*. [online] Available at: <http://dx.doi.org/10.1080/02255189.2016.1197825> [Accessed 20 February 2020].

Bachriadi, D. and Suryana, E., 2020. *Land Grabbing and Speculation for Energy Business: A Case Study of Exxonmobil in East Java, Indonesia*. [online] Available at: https://doi.org/10.1080/02255189.2016.1197825 [Accessed 20 February 2020].

Bachriadi, D., Safitri, H., Sujiwo, T., Suryana, E. and Walhi, A., n.d. *Infrastruktur, Ladang Minyak, Dan Wisata Hijau:Pencaplokan Tanah Di Jawa Pasca Reformasi*. Bandung: Agrarian Resource Centre.

Bachtiar, A., 2016. *Buku Putih: Eksplorasi Indonesia*. Jakarta: Ministry of Energy and Mineral Resources Republic of Indonesia.

Bachtiar, A., 2018. *A Focus Group Discussion on CBM Industry in Indonesia*.

Balfas, R., 2012. Successful CBM Coring Activity Integrated with Conventional Well Drilling in Sanga-sanga Block, Mahakam Delta, East Kalimantan, Indonesia. In: *Indonesian Petroleum Association*. Jakarta: Indonesian Petroleum Association.

Bank Indonesia, 2020. *Data Inflasi – Bank Sentral Republik Indonesia*. [online] Bi.go.id. Available at: https://www.bi.go.id/id/moneter/inflasi/data/Default.aspx [Accessed 14 February 2020].

Bank, G., and Kuuskraa, V., 2006. *The Economics of Powder River Basin Coalbed Methane Development*. Washington D.C.: Advanced Resources International, Inc.

Bayar, Y., 2017. Greenfield and Brownfield Investments and Economic Growth: Evidence from Central and Eastern European Union Countries. *Naše gospodarstvo/Our economy*, 63(3), pp. 19–26.

BHP Billiton, 2019. *Review of the Petroleum Resources Rent Tax BHP Billiton Submission*. [online] BHP Billiton. Available at: https://www.bhp.com/-/media/documents/media/reports-and-presentations/2017/170214_

reviewofthepetroleumresourcesrenttaxbhpbillitonsubmission.pdf? [Accessed 19 March 2020].

Boger, C., Marshall, J. and Pilcher, R., 2014. Worldwide Coal Mine Methane and Coalbed Methane Activities. In: P. Thakur, S. Schatzel and K. Aminian, ed., *Coal Bed Methane: From Prospect to Pipeline*, 1st ed. United States of America: Elsevier Inc., pp. 351–407.

Boone, H. and Boone, D., 2012. Analyzing Likert Data. *Journal of Extension*, [online] 50(2). Available at: https://www.joe.org/joe/2012april/tt2.php [Accessed 8 October 2018].

BOSSERT, T., SOEBEKTI, R. and RAI, N., 1991. 'Bottom-up' planning in Indonesia: Decentralization in the ministry of health. *Health Policy and Planning*, 6(1), pp. 55–63.

Bowe, M. and Moore, T., 2015. Coalbed Methane Potential and Current Realisation in Indonesia. *AAPG*, [online] 90234. Available at: http://www.searchanddiscovery.com/abstracts/pdf/2015/90234gtw/abstracts/ndx_bowe.pdf [Accessed 20 February 2017].

bp.com. 2016. *Statistical Review of World Energy 2015*. [online] Available at: http://www.bp.com/statisticalreview [Accessed 20 October 2016].

BPMigas, 2010a. *CBM Cost Estimation in Indonesia*. Jakarta: BPMigas.

BPMigas, 2010b. *Type of POD and Approval Body*. Jakarta: BPMigas.

Bradshaw, M., Hall, L., Copeland, A. and Hitchins, N., 2012. *Australian Gas Resource Assessment*. Canberra: Geoscience Australia.

Brasier, K., McLaughlin, D., Rhubart, D., Stedman, R., Filteau, M. and Jacquet, J., 2013. Research Articles: Risk Perceptions of Natural Gas Development in the Marcellus Shale. *Environmental Practice*, 15(2), pp. 108–122.

Bria, E., Heller, P., Lassourd, T., Westenberg, E., George-Wagner, M. and Bauer, A., 2016. *Indonesia's Oil and Gas Legislation: Critical Issues*. [ebook] Natural Resources Governance Institute. Available at: https://resourcegovernance.org/sites/default/files/documents/indonesiasoilandgaslegislationcriticalissuesbriefingweb.pdf [Accessed 22 February 2020].

Bryner, G., 2002. *Coalbed Methane Development in the Intermountain West: Primer*. Colorado: University of Colorado Law School.

Busch, T. and Friede, G., 2018. The Robustness of the Corporate Social and Financial Performance Relation: A Second-Order Meta-Analysis. *Corporate Social Responsibility and Environmental Management*, 25(4), pp. 583–608.

Cachia, C. and Lyons, R., 2012. *NSW Ends Royalty Holiday for Petroleum Producers and Increases Penalties*. [online] Mondaq.com. Available at: https://

www.mondaq.com/australia/Energy-and-Natural-Resources/205414/NSW-ends-royalty-holiday-for-petroleum-producers-and-increases-penalties [Accessed 9 February 2020].

Calder, J., 2014. *Administering Fiscal Regimes for Extractive Industries*. [ebook] International Monetary Fund. Available at: https://www.elibrary.imf.org/view/IMF071/20884-9781475575170/20884-9781475575170/20884-9781475575170.xml?language=en&redirect=true [Accessed 21 March 2020].

Callaghan, M., 2017. *Petroleum Resource Rent Tax Review*. Canberra: Australian Government.

Cassidy, N. and Kosev, M., 2015. Australia and the Global LNG Market. RESERVE BANK OF AUSTRALIA, [online] pp. 33–43. Available at: https://www.rba.gov.au/publications/bulletin/2015/mar/pdf/bu-0315-4.pdf [Accessed 13 July 2021].

Cawood, F., 2004. New South African Mineral and Petroleum Resources Development Act: Shaping a New Investment Environment to Reflect a Diverse Society. *Journal of Energy & Natural Resources Law*, 22(2), pp. 121–147.

Cbmasia.ca. 2016. *CBM ASIA*. [online] Available at: http://www.cbmasia.ca/CBM-In-Indonesia# [Accessed 12 April 2016].

Chakhmakhchev, A. and Fryklund, B., 2008. Critical success factors of CBM development – Implications of two strategies to global development. In: *19th World Petroleum Congress*. USA: IHS.

Chatham House the Royal Institute of International Affairs, 2015. *Guidelines for Good Governance in Emerging Oil and Gas Producers*. Energy, Environment and Resources. London: Chatham House The Royal Institute of International Affairs.

Clarke Energy, 2020. *Coal Bed Methane – A Subterranean Energy*. [online] Clarke Energy. Available at: https://www.clarke-energy.com/2011/coal-bed-methane-a-subterranean-energy-treasure/ [Accessed 30 July 2020].

Cockcroft, P., 2008. *Coal Bed Methane in Indonesia*.

Collier, P., 2010. Principles of resource taxation for low-income countries. In: P. Daniel, M. Keen and C. McPherson, ed., *The Taxation of Petroleum and Minerals: Principles, Problem, and Practice*. Oxon: Routledge, pp. 75–86.

Corbin, J. and Strauss, A., 1990. Grounded Theory Research: Procedures, Canons and Evaluative Criteria. *Zeitschrift für Soziologie*, 19(6).

Corbin, J. and Strauss, A., 2008. Basics of Qualitative Research: Techniques and Procedures for Developing Grounded Theory (3rd ed.). Thousand Oaks, CA: Sage. *Organizational Research Methods*, 12(3), pp. 614–617.

Crotty, M., 1998. *The Foundations of Social Research: Meaning and Perspective in the Research Process*. 1st ed. London [i pozostałe]: Sage.

Cully, M., 2015. *Review of the Socioeconomic Impacts of Coal Seam Gas in Queensland*. Canberra: Australian Government, Department of Industry, Innovation, and Science.

Daniel Johnston & Co., Inc, 2018. *Wyoming: Legal and Fiscal Frameworks: Best Practice*. Analysis and Commentary: Fiscal System Study. [online] Available at: http://www.wyia.org/2018/12/03/oil-gas-tax-study-legal-fiscal-frameworks-best-practices/ [Accessed 1 May 2019].

Daniel, P., Keen, M. and McPherson, C., 2010. *The Taxation of Petroleum and Minerals*. London: Routledge/International Monetary Fund.

Davidson, S., 1992. Some Aspects of the Transient Testing of Bowen Basin Coal Seams. In: B. Beamish and P. Gamson, ed., *Symposium on Coal Bed Methane Research and Development in Australia*, 2nd ed. Australia: James Cook University of North Queensland, pp. 153–169.

Davis, R., Noon, S. and Harrington, J., 2007. The petroleum potential of Tertiary coals from Western Indonesia: Relationship to mire type and sequence stratigraphic setting. *International Journal of Coal Geology*, 70(1–3), pp. 35–52.

Deloitte, 2013. *Oil and Gas Taxation in Australia*. Deloitte taxation and investment guides. Deloitte.

Dewan Energi National Indonesia (National Energy Council of Indonesia), 2020. *National Energy Mix 2020*. Jakarta: National Energy Council – Dewan Energi Nasional Indonesia.

Directorate General New and Renewable Energy, 2016. *Buku Statistik EBTKE (Statistical Book of New and Renewable Energy)*. Jakarta: Ministry of Energy and Mineral Resources.

Directorate General of Electricity, 2017. *Electricity Infrastructure Map*. Jakarta: Ministry of Energy and Mineral Resources Republic of Indonesia.

Directorate General of Oil and Gas, 2011. *Focus Group Discussion With Directorate General of Oil and Gas*. Jakarta: Ministry of Energy and Mineral Resources.

Doucet, J. and Brown, C., n.d. *Coal Bed Methane: The Past, Present, and Future of Alberta's Most Abundant Non-Traditional Source of Natural Gas*. 43rd ed.

Downe-Wamboldt, B., 1992. Content analysis: Method, applications, and issues. *Health Care for Women International*, 13(3), pp. 313–321.

Downward, P. and Mearman, A., 2007. Retroduction as mixed-methods triangulation in economic research: Reorienting economics into social science. *Cambridge Journal of Economics*, 31(1), pp. 77–99.

Dupont, P. and McLelland, I., 2013. *Report Analyst of Coal Bed Methane Asia Development Corp.* CBM Asia Development Corp, p. 3.

Dutu, R., 2016. Challenges and policies in Indonesia's energy sector. *Energy Policy*, 98, pp. 513–519.

Easterby-Smith, M., Thorpe, R. and Jackson, P., 2015. *Management and Business Research*. 5th ed. Los Angeles [etc.]: Sage.

Easterby-Smith, M., Thorpe, R., Jackson, P. and Jaspersen, L., 2018. *Management & Business Research*. 6th ed. SAGE.

Edward Freeman, R. and Evan, W., 1990. Corporate governance: A stakeholder interpretation. *Journal of Behavioral Economics*, 19(4), pp. 337–359.

Emam, E., 2015. Gas Flaring in Industry: An Overview. *Petroleum & Coal*, [online] Available at: http://www.vurup.sk/petroleum-coal [Accessed 17 October 2017].

Energy Law No. 30, 2007. No. 30.

Ernst & Young, 2016. *Global Oil and Gas Tax Guide*. Ernst & Young.

Ernst and Young, 2015. *Global Oil and Gas Tax Guide*. Ernst and Young, pp. 27–41.

Esterle, J., Williams, R., Sliwa, R. and Malone, M., 2006. Variability in Coal Seam Gas Content that Impact on Fugitive Gas Emissions Estimations for Australian Black Coals. In: *36th Sydney Basin Symposium 2006:Advances in Study of the Sydney Basin*. Wollongong: University of Wollongong, p. 10.

Evans, D., 2011. *Indonesia's CBM Dream Nears Reality*. [online] Petroleum-economist.com. Available at: https://www.petroleum-economist.com/articles/upstream/exploration-production/2011/indonesia-s-cbm-dream-nears-reality [Accessed 30 July 2020].

Evensen, D., Jacquet, J., Clarke, C. and Stedman, R., 2014. What's the 'fracking' problem? One word can't say it all. *The Extractive Industries and Society*, 1(2), pp. 130–136.

Fauzi, A. and Anna, Z., 2013. The complexity of the institution of payment for environmental services: A case study of two Indonesian PES schemes. *Ecosystem Services*, 6, pp. 54–63.

Fisher, J., 2016. *Environmental Issues and Challenges in Coai Bed Methane Production*. [online] Available at: https://pdfs.semanticscholar.org/8f1d/347a2474068331045353a6576656b42a8bc4.pdf [Accessed 8 June 2017].

Flores, R., 2004a. *Coalbed Methane in the Powder River Basin, Wyoming and Montana: An Assessment of the Tertiary-Upper Cretaceous Coalbed Methane Total Petroleum System*. U.S. Geological Survey Digital Data Series DDS–69–C. [online] Denver: U.S. Geological Survey, p. 32. Available at: https://pubs.usgs.gov/dds/dds-069/dds-069-c/REPORTS/Chapter_2.pdf [Accessed 4 November 2016].

Flores, R., 2004b. Wyodak-Anderson Coal Zone in the Powder River Basin, Montana and Wyoming – A tale of uncorrelatable coalbeds. In: W. Miller, ed., *Coalbed methane and the tertiary geology of the Powder River Basin, Wyoming and Montana*. Wyoming: Wyoming Geological Association, pp. 1–24.

Flores, R., 2014a. *Coal and Coal Bed Gas: Fueling the Future*. United States of America: Elsevier Inc., pp. 567–574.

Flores, R., 2014b. *Coal and Coalbed Gas*. Amsterdam: Elsevier, pp. 2–39.

Freeman, R., 1984. *Strategic Management: A Stakeholder Approach*. Boston: Pitman.

Freeman, R., 2010. *Strategic Management: A Stakeholder Approach*. The United States of America: Cambridge University Press.

Freilich, R. and Popowitz, N., 2012. Oil and Gas Fracking: State and Federal Regulation Does Not Preempt Needed Local Government Regulation. *The Urban Law*, 44(3), pp. 534–575.

Frynas, J., 2009. Corporate social responsibility in the oil and gas sector. *The Journal of World Energy Law & Business*, 2(3), pp. 178–195.

Garnaut, R. and Ross, A., 1975. Uncertainty, Risk Aversion and the Taxing of Natural Resource Projects. *The Economic Journal*, 85(338), p. 272.

General Plan of National Electrification, 2019. *National Electricity General Plan Republic of Indonesia*. Jakarta: Ministry of Energy and Mineral Resources.

General Plan of National Energy. 22. 2017. Republic of Indonesia.

Geo Deltasains Andalan, 2017. *Discussion on Sampling Selection*.

Geospatial Information Agency of Indonesia, 2017. *Beranda*. [online] Big.go.id. Available at: <https://www.big.go.id/> [Accessed 26 September 2020].

Glaser, B. and Strauss, A., 1967. *Discovery of Grounded Theory: Strategies for Qualitative Research*. The United States of America: Aldine Transaction.

Glaser, B., 1978. *Theoretical Sensitivity*. Mill Valley, Calif.: The Sociology Press.

Godfrey, P., Ee, T. and Hewitt, T., 2010. Coal Bed Methane Development in Indonesia: Golden Opportunity or Impossible Dream?. *Journal of Energy & Natural Resources Law*, 28(2), p. 234.

Goerold, W., 2002. Powder River Basin Coalbed Methane Financial Model (PRB-CBM-FM). In: *Coal Bed Methane Development in the Intermountain West*. [online] Colorado: The University of Colorado Law School. Available at: https://scholar.law.colorado.edu/cgi/viewcontent.cgi?referer=https://www.bing.com/&httpsredir=1&article=1011&context=coalbed-methane-development-intermountain-west [Accessed 6 January 2018].

Golubchenko, N., 2018. *Views on CBM Industry in Indonesia (Global Consultancy's Perspective)*.

Government Regulation (GR) 79 (2014): National Energy Policy. No. 79, Article 6.

Government Regulation No 30, 2007.

Government Regulation No. 34, 2005. No. 34, Article 8.

Government Regulation: Upstream Oil and Gas Activities (2004). No. 35.

Graham, A., 2014. *Unconventional Upside*. Jakarta: Risco Energy.

Grote, M., 2007. *Tax Aspects of Domestic Resource Mobilisation – A Discussion of Enduring and Emerging Issues*.

Guj, P., 2012. *Mineral Royalties and Other Mining- Specific Taxes*. Crawley, Western Australia: Australian AID.

Gunawan, B., 2018. *Views on CBM Industry in Indonesia (Government's Perspective)*.

Gylfason, T. and Zoega, G., 2006. Natural Resources and Economic Growth: The Role of Investment. *The World Economy*, 29(8), pp. 1091–1115.

Haldar, S., 2018. Elements of Mining. *Mineral Exploration*, pp. 229–258.

Halstead, J. and Deller, S., 1997. Public Infrastructure in Economic Development and Growth: Evidence from Rural Manufacturers. *Community Development Society. Journal*, 28(2), pp. 149–169.

Hamzah, S., 2018. *A Focus Group Discussion on CBM Industry in Indonesia*.

Harrington, J., 2016. CBM Indonesia – Dull Past, Bright Future. In: *2016 Technical Symposium*. Jakarta: Indonesia Petroleum Association.

Hartiniati, H., 2016. DAMPAK LINGKUNGAN DAN SOSIAL DARI PENGEMBANGAN CBM DI INDONESIA. *Jurnal Teknologi Lingkungan*, 12(2), p. 207.

Heffron, R. and McCauley, D., 2017. The concept of energy justice across the disciplines. *Energy Policy*, 105, pp. 658–667.

Heffron, R. and McCauley, D., 2018. What is the 'Just Transition'?. *Geoforum*, 88, pp. 74–77.

Heffron, R., McCauley, D. and de Rubens, G., 2018. Balancing the energy trilemma through the Energy Justice Metric. *Applied Energy*, 229, pp. 1191–1201.

Horsley & Witten, Inc, 2001. *Draft Evaluation of Impacts to Underground Sources of Drinking Water by Hydraulic Fracturing of Coalbed Methane Reservoirs*. Prepared for the U.S. EPA. EPA-HQ-OW-2004-0032-2543.

HUBERT, J., BUTERA, J. and RICE, R., 1972. Sedimentology of Upper Cretaceous Cody-Parkman Delta, Southwestern Powder River Basin, Wyoming. *Geological Society of America Bulletin*, 83(6), p. 1649.

Hulme, D., 2002. *Coalbed Methane Development in Wyoming's Powder River Basin*. [ebook] Colorado: University of Colorado Law School. Available at: http://scholar.law.colorado.edu/coalbed-methane-development-intermountain-west/4 [Accessed 19 March 2020].

Husen, D., 2008. *ION EXCHANGE & DEIONIZATION: TREATING COAL BED METHANE PRODUCED WATER*. [online] www.waterworld.com. Available at: https://www.waterworld.com/articles/iww/print/volume-8/issue-5/features/ion-exchange-amp-deionization-treating-coal-bed-methane-produced-water.html [Accessed 6 December 2018].

Husin, M., 2018. *Views on CBM Industry in Indonesia (Investor's Perspective)*.

Hutton, A., 2009. Geological Setting of Australasian Coal Deposits. In: R. Kininmonth and E. Baafi, ed., *Australasian Coal Mining Practice*. Victoria: The Australasian Institute of Metallurgy, pp. 40–84.

Ialnazov, D. and Keeley, A., 2020. Motivations, Enabling Factors and Barriers to the Energy Transition in Indonesia and Vietnam. *IOP Conference Series: Earth and Environmental Science*, 505, p. 012044.

ICF International, 2014. *Recent Australian Natural Gas Pricing Dynamics and Implications for the U.S. LNG Export Debate*. ICF Resources.

IESR, 2019. *Indonesia Clean Energy Outlook: Tracking Progress and Review of Clean Energy Development in Indonesia*. Jakarta: Institute for Essential Services Reform (IESR).

Indonesia Government's Law & Regulation, 2001. *Oil and Gas Law*. No. 22. Jakarta: Government of Indonesia.

Indonesia Government's Law & Regulation, 2004. *Guidance and Supervision of Upstream Business Activities*. No. 35. Jakarta: Ministry of Energy and Mineral Resources.

Indonesia Government's Law & Regulation, 2010. *Cost Recovery and Indonesia's Oil and Gas Taxation*. No. 79. Jakarta: Ministry of Finance.

Indonesian Fiscal Policy Office, 2008. *Low Carbon Development Options for Indonesia*. Jakarta: Indonesian Fiscal Policy Office.

Indonesian Petroleum Association, 2013. *Speech on Indonesian Petroleum Association Convention and Exhibition*. Jakarta.

Indonesian Petroleum Association, 2016. *Oil and Gas Sector to Be the Motor of The Economy*.

Inter-Governmental Forum – OECD, 2020. *The Hidden Cost of Tax Incentives in Mining*. [online] IGF – OECD. Available at: https://www.oecd.org/tax/beps/hidden-cost-of-tax-incentives-in-mining-draft-toolkit-oecd-igf.pdf [Accessed 20 March 2020].

International Energy Agency, 2014. *Energy Supply Security*. Indonesia. Paris: International Energy Agency, p. 22.

International Energy Agency, 2019. *The Role of Gas in Today's Energy Transitions*. World Energy Outlook Special Report. [online] Paris: International Energy Agency. [Accessed 13 May 2020].

International Energy Agency, 2020. *The Role of Gas in Today's Energy Transitions.* [online] Paris: International Energy Agency. Available at: https://webstore.iea.org/download/direct/2819?fileName=TheRoleofGas.pdf [Accessed 10 November 2020].

International Monetary Fund, 2012. *Fiscal Regimes for Extractive Industries: Design and Implementation.* [online] International Monetary Fund. Available at: https://www.google.com/url?sa=t&rct=j&q=&esrc=s&source=web&cd=3&cad=rja&uact=8&ved=2ahUKEwi6nozFq9HoAhWxQkEAHVs5BTQQFjACegQIARAB&url=https%3A%2F%2Fwww.imf.org%2Fexternal%2Fnp%2Fpp%2Feng%2F2012%2F081512.pdf&usg=AOvVaw3%2D%2DNziWcwfWOVklkY88BOl [Accessed 20 March 2020].

International Monetary Fund, 2015. *"Options for Low Income Countries' Effective and Efficient Use of Tax Incentives for Investment."* [online] International Monetary Fund. Available at: https://www.imf.org/external/np/g20/pdf/101515.pdf [Accessed 20 March 2020].

Investments, I., 2015. *Natural Gas in Indonesia – Indonesian LPG Production & Export | Indonesia Investments.* [online] Indonesia-investments.com. Available at: http://www.indonesia-investments.com/business/commodities/natural-gas/item184 [Accessed 7 March 2016].

Investments, I., 2016. *Crude Oil Indonesia – Indonesian Oil Production & Consumption | Indonesia Investments.* [online] Indonesia-investments.com. Available at: http://www.indonesia-investments.com/business/commodities/crude-oil/item267 [Accessed 7 March 2016].

IPCC, n.d. IPCC Special Report on Carbon dioxide Capture and Storage. [online] IPCC, pp. 384–400. Available at: https://www.ipcc.ch/site/assets/uploads/2018/03/srccs_annex1-1.pdf [Accessed 13 July 2021].

IRENA, 2017. Accelerating the Energy Transition through Innovation. WORKING PAPER based on global REmap analysis. [online] Abu Dhabi: IRENA. Available at: https://irena.org/publications/2017/Jun/Accelerating-the-Energy-Transition-through-Innovation [Accessed 13 July 2021].

IRENA, 2020. *Energy Transition.* [online] Irena.org. Available at: https://www.irena.org/energytransition [Accessed 20 July 2020].

Jackson, H., n.d. *Drilling for Tax Credits: Runaway Coalbed Methane Development Doesn't Need a Tax Break.* [ebook] Public Citizen. Available at: https://www.citizen.org/sites/default/files/coalbedfactsheet.pdf [Accessed 24 April 2019].

Jacquet, J. and Stedman, R., 2013. Perceived Impacts from Wind Farm and Natural Gas Development in Northern Pennsylvania. *Rural Sociology*, 78(4), pp. 450–472.

Jacquet, J., 2012. Landowner attitudes toward natural gas and wind farm development in northern Pennsylvania. *Energy Policy*, 50, pp. 677–688.

Jalbert, K., Kinchy, A. and Perry, S., 2013. Civil society research and Marcellus Shale natural gas development: Results of a survey of volunteer water monitoring organizations. *Journal of Environmental Studies and Sciences*, 4(1), pp. 78–86.

Jalbert, K., Kinchy, A. and Perry, S., 2014. Civil society research and Marcellus Shale natural gas development: results of a survey of volunteer water monitoring organizations. *Journal of Environmental Studies and Sciences*, 4(1), pp. 78–86.

Jenkins, C. and Boyer, C., 2016. *Coalbed- and Shale-Gas Reservoirs*. [online] www.onepetro.org. Available at: https://www.onepetro.org/journal-paper/SPE-103514-JPT [Accessed 29 October 2016].

Jenkins, K., McCauley, D., Heffron, R., Stephan, H. and Rehner, R., 2016. Energy justice: A conceptual review. *Energy Research & Social Science*, 11, pp. 174–182.

Johnson, C., 1990. Ranking countries for minerals exploration. *Natural Resources Forum*, 14(3), pp. 178–186.

Johnston, D. and Johnston, D., 2018. *Petroleum Fiscal Regimes: Production Sharing Contract*.

Johnston, D., 1994. *International Petroleum Fiscal Systems and Production Sharing Contracts*. 1st ed. Tulsa: PennWell Publishing Company.

Johnston. D., 2018. *Wyoming: Legal and Fiscal Frameworks: Best Practice. Analysis and Commentary: Fiscal System Study*. [online] Available at: <http://www.wyia.org/2018/12/03/oil-gas-tax-study-legal-fiscal-frameworks-best-practices/> [Accessed 1 May 2019].

Johnston, D., 2019. *Severance & Property Tax & Ad Valorem Tax*. [email].

Kalahari Energy Botswana, 2020. *How Is Coal Bed Methane Used – Kalahari Energy Botswana*. [online] Kalahari Energy Botswana. Available at: https://keb.co.bw/how-is-cbm-used// [Accessed 30 July 2020].

Kalsel.bps.go.id. 2019. *Badan Pusat Statistik Provinsi Kalimantan Selatan*. [online] Available at: https://kalsel.bps.go.id/statictable/2016/10/10/689/jumlah-penduduk-kalimantan-selatan-menurut-kabupaten-kota-jenis-kelamin-dan-kepadatan-penduduk-2010-hasil-sp2010-angka-final-.html [Accessed 18 June 2019].

Kasatuka, C. and Minnitt, R., 2006. Investment and non-commercial risks in developing countries. *The Journal of the Southern African Institute of Mining and Metallurgy*, 106(12), pp. 849–856.

Kasriel, K. and Wood, D., 2013. *Upstream Petroleum Fiscal and Valuation Modelling in Excel.* West Sussex: John Wiley & Sons Ltd.

Kemenkeu.go.id. 2020. *Ministry of Finance, Republic of Indonesia.* [online] Available at: https://www.kemenkeu.go.id/publikasi/berita/ini-komitmen-pemerintah-kurangi-emisi-karbon/ [Accessed 30 July 2020].

Kennedy, S., 2018. Indonesia's energy transition and its contradictions: Emerging geographies of energy and finance. *Energy Research & Social Science*, 41, pp. 230–237.

Khan, M., 2000. Rents, Efficiency and Growth. In: M. Khan and K. Jomo, ed., *Rents, Rent-Seeking and Economic Development.* Cambridge: Cambridge University Press.

Kondracki, N., Wellman, N. and Amundson, D., 2002. Content Analysis: Review of Methods and Their Applications in Nutrition Education. *Journal of Nutrition Education and Behavior*, 34(4), pp. 224–230.

Kraal, D., 2013. A Grounded Theory Approach to the Minerals Resource Rent Tax. *SSRN Electronic Journal*, (4), pp. 841–874.

Kraal, D., 2016. *The Petroleum Resource Rent Tax: Overview of Primary Documents and Literature Leading to the 1987 Legislation.* TTPI – Working Paper 9/2016 October 2016. Canberra: The Australian National University.

Kraal, D., 2017. Review of Australia's Petroleum Resource Rent Tax: Implications from a Case Study of the Gorgon Gas Project. *Federal Law Review*, 45(2), pp. 315–349.

Kriesky, J., Goldstein, B., Zell, K. and Beach, S., 2013. Differing opinions about natural gas drilling in two adjacent counties with different levels of drilling activity. *Energy Policy*, 58, pp. 228–236.

Kristiansen, S. and Pratikno, 2006. Decentralising education in Indonesia. *International Journal of Educational Development*, 26(5), pp. 513–531.

Kukarkab.bps.go.id. 2019. *Badan Pusat Statistik Kabupaten Kutai Kartanegara.* [online] Available at: https://kukarkab.bps.go.id/statictable/2016/09/15/16/jumlah-penduduk-dan-rasio-jenis-kelamin-menurut-kecamatan-di-kabupaten-kutai-kartanegara-2015.html [Accessed 18 June 2019].

Kunce, M., Gerking, S., Morgan, W. and Maddux, R., 2003. State Taxation, Exploration, and Production in the U.S. Oil Industry. *Journal of Regional Science*, 43(4), pp. 749–770.

Kusumah, G. and Munawar, A., 2015. How to Make Coal Bed Methane Work in Indonesia. In: *Indonesian Petroleum Association Conference.* Jakarta: Association for American Petroleum Geologist.

Kuuskraa, V. and Boyer, C., 1993. Economic and Parametric Analysis of Coal Bed Methane. In: *Hydrocarbons for Coal.* Tulsa, OK, USA: The American Association of Petroleum Geologist, pp. 373–394.

Ladd, A., 2013. *Stakeholder Perceptions of Socioenvironmental Impacts for Unconventional Natural Gas Development and Hydraulic Fracturing in the Haynesville Shale*. Texas: Southern Rural Sociological Association, pp. 56–89.

Lalean, B., 2008. Designing a Production Sharing Contract (PSC) for Coal Bed Methane Project in order to Balance the Shares Portions of the Government and the Contractor. In: *Indonesian Petroleum Association Conference*. Jakarta: Association for American Petroleum Geologist.

Lalean, B., 2010. Launching a First Coal Bed Methane Project in Indonesia – A Case Study. In: *Indonesian Petroleum Association Conference*. Jakarta: Association for American Petroleum Geologist.

Lalean, B., 2012. Seeking Workable CBM Business Model in Indonesia. In: *Indonesian Petroleum Association Conference*. Jakarta: Association for American Petroleum Geologist.

Land, D. and Jones, C., 1987. Coal geology and exploration of part of the Tertiary Kutei Basin in East Kalimantan, Indonesia. *Geological Society, London, Special Publications*, 32(1), pp. 235–255.

Law on Land Agrarian No. 2, 2012. No. 2.

Leedy, P. and Ormrod, J., 2001. *Practical Research*. Upper Saddle River, N.J.: Merrill Prentice Hall.

Leeuw, E., Hox, J. and Dillman, D., 2008. *International Handbook of Survey Methodology*. 1st ed. New York: Lawrence Erlbaum Associates.

Legowo, E., 2018. *Views on CBM Industry in Indonesia (Government's Perspective)*.

Len, C., 2014. *Indonesia Coalbed Methane*. Bountiful Promise Help Up by Institutional Paralysis. [online] Singapore: Energy Studies Institute. Available at: http://esi.nus.edu.sg/docs/default-source/esi-policy-briefs/indonesia-cbm-bountiful-promise-held-up-by-institutional-paralysis.pdf [Accessed 28 March 2016].

Letts, L., 2012. *Coal Seam Gas Production – Friend Or Foe of Queensland's Water Resources?*. 1st ed. Australia: Thomson Reuters.

Li, S. and Zhang, B., 2016. Research of Coalbed Methane Development Well-Type Optimization Method Based on Unit Technical Cost. *Sustainability*, 8(9), p. 843.

Likert, R., 1974. The Method of Constructing an Attitude Scale. In: G. Maranell, ed., *Scaling: A Source book for Behavioral Scientists*, 1st ed. Illinois: Aldine Publishing Company, pp. 233–243.

Loe, J. and Kelman, I., 2020. *Arctic Petroleum's Community Impacts: Local Perceptions for Hammerfest, Norway*.

Lubiantara, B., 2012. *Ekonomi Migas: Tinjauan Aspek Komersial Kontrak Migas*. Jakarta: Grasindo.

Lubiantara, B., 2017. *Paradigma Baru: Pengelolaan Sektor Hulu Migas Dan Ketahanan Energi – New Paradigm: Managing the Upstream Industry and Energy Security*. Jakarta: Grasindo.

Lubiantara, B., 2018. *Views on CBM Industry in Indonesia (Local Consultancy's & Government's Perspective)*.

MacKinnon, A. and Fox, K., 2006. Demanding Beneficial Use: Opportunities and Obligations for Wyoming Regulators in Coalbed Methane. *Wyoming Law Review*, [online] 6(2), pp. 369–400. Available at: https://scholarship.law.uwyo.edu/cgi/viewcontent.cgi?article=1135&context=wlr [Accessed 19 March 2020].

Mair, D., 2018. *Views on CBM Industry in Indonesia (Global Consultancy's Perspective)*.

Makmun and Sitepu, E., 2012. Makmun. *Effect of fiscal incentives on coal bed methane price: a hypothetical analysis*, 3(1).

Mallet, C., Pattison, C., McLennan, T., Balve, P. and Sullivan, D., 2020. Bowen Basin. In: C. Ward, H. Harrington, C. Mallet and I. Beeston, ed., *Geology of Australian Coal Basin*, 1st ed. Australia: Geological Society of Australia, pp. 299–339.

Mallick, N. and Prabu, V., 2017. Energy analysis on Coalbed Methane (CBM) coupled power systems. *Journal of CO_2 Utilization*, 19, pp. 16–27.

Mann, H. and Whitney, D., 1947. On a Test of Whether one of Two Random Variables is Stochastically Larger than the Other. *The Annals of Mathematical Statistics*, 18(1), pp. 50–60.

Marquardt, J., 2014. A Struggle of Multi-level Governance: Promoting Renewable Energy in Indonesia. *Energy Procedia*, 58, pp. 87–94.

Mashari, D. and Sumandra, M., 2019. Analysis on the Implementation of Gross Split Production Sharing Contract: Simulation on the Oil and Gas' Project Economics in Indonesia. In: *United States Association for Energy Economics (USAEE)*. Jakarta: The Special Task Force for Upstream Oil and Gas Business Activities (SKK Migas).

Massarotto, P., 1999. Cost Benefit Analysis of Coalbed Methane Recovery Activities in Australia and New Zealand- Implications for Commercial Projects and Government Policy. In: M. Mastalerz, M. Glikson and S. Golding, ed., *Coalbed Methane: Scientific, Environmental and Economic Evaluation*. Dordrecht: Kluwer Academic Publishers, pp. 33–54.

Mastalerz, M., 2014. Coal Bed Methane. *Future Energy*, pp. 145–158.

Mayer, A., 2016. Risk and benefits in a fracking boom: Evidence from Colorado. *The Extractive Industries and Society*, 3(3), pp. 744–753.

McCauley, D., Ramasar, V., Heffron, R., Sovacool, B., Mebratu, D. and Mundaca, L., 2019. Energy justice in the transition to low carbon energy systems: Exploring key themes in interdisciplinary research. *Applied Energy*, 233–234, pp. 916–921.

McLaren, J. and Chabal, P., 2011. Given the Fact That Australia has had a 'Petroleum Resource Rent Tax' Since 1987, Why Should There Be Any Opposition to a 'Mineral Resource Rent Tax' ?. *Journal of the Australasian Tax Teachers Association*, 6(1).

Meng, Q., 2017. The impacts of fracking on the environment: A total environmental study paradigm. *Science of the Total Environment*, 580, pp. 953–957.

Meredith, E., Wheaton, J., Kuzara, S., Donato, T., Bierbach, S. and Schwarz, C., 2010. *2009 Annual Coalbed Methane Regional Ground Water Monitoring Report: Northern Portion of the Powder River Basin, Montana*. Open File Report 591. Montana: Montana Bureau of Mines and Energy.

Metters, C., 2014. Delivering Value in PSC Environment. In: *Indonesian Petroleum Association, 38th Annual Convention & Exhibition*. Jakarta: Indonesian Petroleum Association.

Mian, M., 2011. *Project Economics and Decision Analysis*. Tulsa, Okla.: PennWell Corp.

Ministry Decree: GHG Emissions Standard (2003). No. 29.

Ministry of Energy and Mineral Resources, 2005. *Marginal Field Development Incentives*. No. 008. Jakarta: Ministry of Energy and Mineral Resources.

Ministry of Energy and Mineral Resources, 2008a. *No. 36 Accelerating Coal Bed Methane in Indonesia*. Jakarta: MoEMR.

Ministry of Energy and Mineral Resources, 2008b. *The Coal Bed Methane Activities*. No. 036. Jakarta: Ministry of Energy and Mineral Resources.

Ministry of Energy and Mineral Resources, 2008c. *The Cost Recovery Items*. No. 022. Jakarta: Ministry of Energy and Mineral Resources.

Ministry of Energy and Mineral Resources, 2014. *Conditions and Procedure for Allocation and Utilization As Well As Natural Gas Price*. No. 37. Jakarta: Ministry of Energy and Mineral Resources.

Ministry of Energy and Mineral Resources, 2015. *The Coal Bed Methane Acceleration*. No. 038. Jakarta: Ministry of Energy and Mineral Resources.

Ministry of Energy and Mineral Resources, 2021. Intelligent Strategies Power and Utility Sector to Achieve Indonesia's Carbon Neutral by 2050.

Ministry of Legal and Human Rights, Republic of Indonesia, 2007. *Constitution Law on Limited Company, No. 40*. Jakarta: Ministry of Law and Human Rights, Republic of Indonesia.

Ministry Regulation (2012): The Procedures for Determining and Tendering Unconventional Working Contract Areas. No. 05.

Ministry Regulation No. 30, 2017.

Ministry Regulation on Local Participating Interest No 37, 2016. No. 37.

Ministry Regulation: Environmental Impacts and Sustainability (2012). 5.

Ministry Regulation: Fiscal Regimes for CBM in Indonesia (2015). No. 38.

Mintz, J. and Chen, D., 2012. Capturing Economic Rents from Resources Through Royalties and Taxes. *SSRN Electronic Journal*,.

Moektianto, S., 2018. *Views on CBM Industry in Indonesia (Investor's and Industry Association's Perspective)*.

Mohamad, M., 2015. *Progressing CBM Development in Indonesia*.

Mommer, B., 2000. *The Governance of International Oil: The Changing Rules of the Game*. 1st ed. Oxford: Oxford Institute for Energy Studies.

Montoya, D., 2012. *Coal Seam Gas Royalties in Australian States & Territories*. New South Wales Parliament.

Mooij, R. and Klemm, A., 2021. Why and How to Tax Corporate Income. In: R. Mooij, A. Klemm and V. Perry, ed., Corporate Income Taxes under Pressure. [online] International Monetary Fund, p. 388. Available at: https://www.elibrary.imf.org/view/books/071/28329-9781513511771-en/ch002.xml [Accessed 13 July 2021].

Moore, T., 2012. Coal Bed Methane: A Review. *International Journal of Coal Geology*, 101(36–81).

Moore, T., Bowe, M. and Nas, C., 2014. High heat flow effects on a coalbed methane reservoir, East Kalimantan (Borneo), Indonesia. *International Journal of Coal Geology*, 131, pp. 7–31.

Morgan, P., 2002. Mineral title management—the key to attracting foreign mining investment in developing countries?. *Applied Earth Science*, 111(3), pp. 165–170.

Morisset, J., 1999. *Foreign Direct Investment in Africa: Policies Also Matter*. World Bank Policy Research Working Paper, No. 2481. [online] World Bank. Available at: http://papers.ssrn.com/sol3/papers.cfm?abstract_id=632551 [Accessed 20 March 2020].

Moss, S. and Carter, A., 1996. Thermal histories of Tertiary sediments in western central Sumatra, Indonesia. *Journal of Southeast Asian Earth Sciences*, 14(5), pp. 351–371.

Muaraenimkab.bps.go.id. 2019. *BPS Kabupaten Muara Enim*. [online] Available at: https://muaraenimkab.bps.go.id/statictable/2018/09/12/35/jumlah-dan-pertumbuhan-penduduk-di-kabupaten-muara-enim-tahun-2011-2017.html [Accessed 18 June 2019].

Mudd, M., 2012. Montana v. Wyoming: An Opportunity to Right the Course for Coalbed Methane Development and Prior Appropriation. *Golden Gate University Environmental Law Journal*, 5(2), pp. 297–342.

Munawar, A., Hendromurti, D., Dewi, R., Foglio, G. and Schupp, B., 2013. Dewatering Facilities for CBM Resource Appraisal, Lesson Learnt from Sanga-sanga CoalBed Methane, East Kalimantan. In: *Indonesian Petroleum Association*. Jakarta: Indonesian Petroleum Association.

National Electricity General Plan Republic of Indonesia, 2019. *National Electricity General Plan Republic of Indonesia*. Jakarta: Ministry of Energy and Mineral Resources.

National Research Council, 2010. Coalbed Methane Produced Water Management and Beneficial Uses. In: National Research Council, ed., *Management and Effects of Coalbed Methane Produced Water in the Western United States*. [online] Washington, DC: The National Academies Press. Available at: https://doi.org/10.17226/12915. [Accessed 6 May 2019].

National Resource Governance Institute, 2015. *Fiscal Regime Design*. What Revenues the Government Will be Entitled to Collect. [online] New York: National Resource Governance Institute. Available at: https://resourcegovernance.org/sites/default/files/nrgi_Fiscal-Regime-Design.pdf [Accessed 9 February 2020].

Natural Resource Governance Institute, 2015. *Fiscal Regime Design: What Revenues the Government Will Be Entitled to Collect*. Natural Resource Governance Institute.

Natural Resources Charter, 2014. *Fiscal Regimes & Contract Terms*. Technical Guide. Natural Resources Charter.

NN, 2015. *CBM Project Economics Calculation*. [Microsoft Excel] Jakarta.

Noor, R., 2015. Analisis Kebijakan Gas Metana Batubara Indonesia Saat Ini: "Analysis on the Current CBM Policy in Indonesia)." *Kompasiana*, [online] Available at: https://www.kompasiana.com/ryanalfian/55009f6ca33311ac0a5100d3/analisis-kebijakan-gas-metana-batubara-indonesia-saat-ini [Accessed 19 February 2020].

Oentoeng Suria & Partners, n.d. *The Challenges of CBM Development in Indonesia*.

OffShore Energy, 2013. *CBM Asia Updates on Operations in Indonesia – Offshore Energy*. [online] Offshore Energy. Available at: <https://www.offshoreenergy.biz/cbm-asia-updates-on-operations-in-indonesia-2/> [Accessed 26 September 2020].

O'Neill, D., 1993. Mining investment in Africa. *Natural Resources Forum*, 17(4), pp. 262–272.

Orlitzky, M., Schmidt, F. and Rynes, S., 2003. Corporate Social and Financial Performance: A Meta-Analysis. *Organization Studies*, 24(3), pp. 403–441.

Otto, J., 1992. *A Global Survey of Mineral Company Investment Preferences. In Mineral Investment Conditions in Selected Countries of the Asia-Pacific Region*. New York: United Nations.

Owusu, V., 2019. Impacts of the petroleum industry on the livelihoods of fisherfolk in Ghana: A case study of the Western Region. *The Extractive Industries and Society*, 6(4), pp. 1256–1264.

Oxford Policy Management Limited, 2015. *Low Carbon Development in Indonesia*. Oxford: Oxford Policy Management Limited.

Pätäri, S., Arminen, H., Tuppura, A. and Jantunen, A., 2014. Competitive and responsible? The relationship between corporate social and financial performance in the energy sector. *Renewable and Sustainable Energy Reviews*, 37, pp. 142–154.

Penney, K., McCallum, R., Schultz, A. and Ball, A., 2007. *Mineral Exploration In APEC Economies: A Framework for Investment*. Australian Bureau of Agricultural and Resource Economics (ABARE) research report, December. [online] Australian Bureau of Agricultural and Resource Economics (ABARE). Available at: http://publications.apec.org/publication-detail.php?pub_id=244 [Accessed 20 March 2020].

Petrowiki.org. 2017. *CBM Production Operations -*. [online] Available at: http://petrowiki.org/CBM_production_operations [Accessed 2 September 2017].

Population.un.org. 2020. *World Population Prospects – Population Division – United Nations*. [online] Available at: https://population.un.org/wpp/Download/Standard/Population/ [Accessed 6 July 2020].

Prayogo, D., 2013. Measuring Corporate Social Responsibility for Local Communities in Mining, Oil and Gas Industries, the Case of Indonesia. *Journal of Economics and Sustainable Development*, 4(1).

President Regulation, 2017. *General Plan of National Energy*. Jakarta: President of Indonesia.

President Regulation: Domestic Gas Price Determination (2016). No. 40.

PricewaterhouseCoopers, 2014. *Oil and Gas in Indonesia*. Investment and Taxation Guide, May 2014 – 6th edition. Jakarta: Pricewaterhouse Coopers.

PricewaterhouseCoopers, 2015. *Oil and Gas in Indonesia*. Investment and Taxation Guide, May 2014 – 6th edition. Jakarta: PricewaterhouseCoopers.

PricewaterhouseCoopers, 2016. *Oil and Gas in Indonesia*. Investment and Taxation Guide, May 2016 – 7th edition. Jakarta: Pricewaterhouse Coopers.

Pujobroto, A., 2018. *Views on CBM Industry in Indonesia (Investor's Perspective).*

Purba, S., 2018. *Views on CBM Industry in Indonesia (Government Perspective , Written Interview).*

Purwanto, W., Muharam, Y., Pratama, Y., Hartono, D., Soedirman, H. and Anindhito, R., 2016. Status and outlook of natural gas industry development in Indonesia. *Journal of Natural Gas Science and Engineering*, 29, pp. 55–65.

Putrohari, R., Kasyanto, A., Suryanto, H. and Rasyid, I., 2007. PSC Term and Condition and its Implementation in South East Asia Region. In: *Indonesian Petroleum Association*. Jakarta.

Rahmadhi, H., 2018. *A Focus Group Discussion on CBM Industry in Indonesia.*

Redhead, A., 2018. *Tax Incentives in Mining: Minimising Risks to Revenue.* [ebook] Canada: International Institute for Sustainable Development. Available at: https://www.google.com/url?sa=t&rct=j&q=&esrc=s&source=web&cd=2&cad=rja&uact=8&ved=2ahUKEwjvmofc-cnoAhWpQxUIHQeZDmYQFjABegQIBBAB&url=https%3A%2F%2Fwww.oecd.org%2Ftax%2Fbeps%2Ftax-incentives-in-mining-minimising-risks-to-revenue-oecd-igf.pdf&usg=AOvVaw02chYk5AZ00FBSPJnbvrf3 [Accessed 20 March 2020].

Reserve Bank of Australia. 2017. *Inflation Calculator | RBA*. [online] Available at: https://www.rba.gov.au/calculator/ [Accessed 6 May 2019].

Riau.bps.go.id. 2019. *Badan Pusat Statistik Provinsi Riau*. [online] Available at: https://riau.bps.go.id/statictable/2017/01/16/233/-jumlah-penduduk-provinsi-riau-menurut-jenis-kelamin-dan-kabupaten-kota-2016.html [Accessed 18 June 2019].

Rice, C. and Nuccio, V., 2000. *Water Produced With Coal-Bed Methane*. [ebook] Denver: U.S. Department of the Interior, U.S. Geological Survey. Available at: https://pubs.usgs.gov/fs/fs-0156-00/fs-0156-00.pdf [Accessed 6 May 2017].

Rightmire, C., 1984. *Coalbed Methane Resources of the United States*. Tulsa: American Association of Petroleum Geologists, pp. 1–13.

Roarty, M., 2011. *The Development of Australia's Coal Seam Gas Resources*. The Parliament of Australia.

Rodvelt, G., 2014. Vertical Well Construction and Hydraulic Fracturing for CBM Completions. *Coal Bed Methane*, pp. 101–135.

Rogan, D., 2014. *Charting Indonesia's Oil, Gas and Mining for Local Extraction to Local Benefits | Extractive Industries Transparency Initiative*. [online] Eiti.org. Available at: https://eiti.org/news/charting-indonesias-oil-gas-mining-from-local-extraction-to-local-benefits [Accessed 6 February 2017].

Rogers, R., Ramurthy, K., Rodvelt, G. and Mullen, M., 2007. *Coalbed Methane: Principles and Practices*. Starkville, Miss.: Halliburton.

Rosyadi, H., 2010. *Potensi Coalbed Methane Untuk Mensukseskan Program Diversifikasi Energi Nasional*. [ebook] Available at: https://www.researchgate.net/publication/282183107_Potensi_Coalbed_Methane_untuk_Mensukseskan_Program_Diversifikasi_Energi_Nasional [Accessed 22 August 2018].

S&P Global, 2020. *What Is Energy Transition?*. [online] Spglobal.com. Available at: https://www.spglobal.com/en/research-insights/articles/what-is-energy-transition [Accessed 20 July 2020].

SAGE, 2012. Corporate Social Responsibility and Corporate Social Performance. In: *Sage Brief Guide to Business Ethics*. [online] Thousand Oaks: Sage Publisher Inc., pp. 231–248. Available at: https://doi.org/10.4135/9781452243979.n26 [Accessed 16 September 2018].

Saghafi, A. and Hadiyanto, 2000. ABSTRACT: Methane reservoir properties of Indonesian coals. *AAPG Bulletin*, 84.

Salant, P. and Dillman, D., 1994. *How to Conduct Your Own Survey*. 1st ed. New York: Wiley.

Saputra, R., 2015. *Preliminary Interview on Plan Development Process*.

Saputra, R., 2017. *Informal Discussion on PSC Issues*.

Schlosberg, D., 2003. The justice of environmental justice: Reconciling equity, recognition, and participation in a political movement. In: A. Light and A. De-Shalit, ed., *Moral and political reasoning in environmental practice*. Cambridge, Mass.: MIT Press.

Schultz, K. and Adler, L., 2016. *Environmental and Sustainability Assessment of Current and Prospective Status of Coal Bed Methane Production and Use in the European Union*. Italy: EU Commission, JRC Science Hub, European Union.

Schwab, K., 2020. *The Global Competitiveness Report 2017–2018*. [ebook] Geneva: World Economic Forum, pp. 148–149. Available at: http://www3.weforum.org/docs/GCR2017-2018/05FullReport/TheGlobalCompetitivenessReport2017–2018.pdf [Accessed 23 February 2020].

Setiawan, E., Maryudi, A., Purwanto, R. and Lele, G., 2016. Opposing interests in the legalization of non-procedural forest conversion to oil palm in Central Kalimantan, Indonesia. *Land Use Policy*, 58, pp. 472–481.

Setyatmoko, U., 2018a. *A Focus Group Discussion on CBM Industry in Indonesia.*

Setyatmoko, U., 2018b. *Views on CBM Industry in Indonesia (Investor's Perspective).*

Sider, A. and Matthews, C., 2017. Henry Hub Emerges as Global Natural Gas Benchmark. *The Wall Street Journal*, [online] Available at: https://www.cmegroup.com/education/files/henry-hub-emerges-as-a-global-gas-benchmark.pdf. [Accessed 10 February 2020].

Sigam, C. and Garcia, L., 2012. *Extractive Industries: Optimizing Value Retention in Host Countries.* [ebook] Geneva: United Nations Conference on Trade and Development. Available at: https://unctad.org/en/PublicationsLibrary/suc2012d1_en.pdf [Accessed 23 March 2020].

Sinner, J. and Scherzer, J., 2014. *The Public Interest in Resource Rent.*

Sirait, D., 2013. *Indonesia Current Policy and Regulation.*

SKKMigas, 2010. *Pedoman Tata Kerja (Guidelines and Work Procedures for Plan of Development) 072.* Jakarta: SKKMigas, p.www.skkmigas.go.id.

SKKMigas, 2015. *Pedoman Tata Kerja (Guidelines and Work Procedures for Accounting Guidelines of Oil and Gas Working Contract) 059.* Jakarta: SKKMigas, p.www.skkmigas.go.id.

SKKMigas, 2017. *Pedoman Tata Kerja (Guidelines and Work Procedures for Good and Service Tender) 007.* Jakarta: SKKMigas, p.www.skkmigas.go.id.

SKKMigas, 2018. *Coal Bed Methane Blocks in Indonesia 2008–2018.* Jakarta: SKKMigas.

Skkmigas.go.id. 2016. *SKKMIGAS » Statistics of Production.* [online] Available at: http://www.skkmigas.go.id/en/statistik/statistik-produksi [Accessed 4 April 2016].

Skov, J. and Myers, N., 2004. *Easy Money, Hidden Costs: Applying Precautionary Economic Analysis to Coalbed Methane in the Powder River Basin.* [online] Iowa: Science and Environmental Health Network. Available at: http://www.sehn.org/tccpdf/coalbed%20methane%20costs%20Powder%20River.pdf [Accessed 16 September 2017].

Smith, J., 2012. *Modelling the Impact of Taxes on Petroleum Exploration and Development.* WP/12/278. International Monetary Fund.

Soeryowibowo, M., 2018. Views on CBM Industry in Indonesia (Investor's and Industry Association's Perspective).

Soetrisno, S., 2018. *Views on CBM Industry in Indonesia (Industry Observer's Perspective).*

Soh, M., 2011. *Australian Coal Bed Methane: Principles and Development Challenges.*

Srinivasu, B. and Rao, P., 2013. Infrastructure Development and Economic growth: Prospects and Perspective. *Journal of Business Management & Social Sciences Research (JBM&SSR)*, 2(1), pp. 81–91.

Stevens, H., Scott, H., Sani, K. and Hardjosuwiryo, S., 2001. Indonesia's 337 TCF CBM Resource a Low Cost Alternative to Gas, LNG. *Oil and Gas Journal*, (October 22).

Stevens, S. and Hadiyanto, 2004. Indonesia: Coalbed Methane Indicators and Basin Evaluation. In: *SPE Asia Pacific Oil and Gas Conference and Exhibition*. Society of Petroleum Engineers.

Steyn, J. 2019. *Hydraulic Fracturing of Rock Formations Part 2: Coal-bed Methane Recovery*. 1st ed. [ebook] Owner Team Consultation. Available at: <https://www.researchgate.net/publication/336441808_Hydraulic_Fracturing_of_Rock_Formations_Part_2_Coal-bed_Methane_Recovery> [Accessed 26 September 2020].

Stickley, D., 2014. *A Framework for Negotiating and Managing Gas Industry Contracts*. 2nd ed. Dundee.

Streimikiene, D., Simanaiciene, Z. and Kovaliov, R., 2009. Corporate Social Responsibility for Implementation of Sustainable Energy Development in Baltic States. *Renewable and Sustainable Energy Reviews*, 13, pp. 813–824.

Sukhyar, R. and Fakhruddin, R., 2013. *Unconventional Oil & Gas Potential in Indonesia With Special Attention to Shale Gas and Coal Bed Methane*.

Sumarno, T., n.d. *CBM Development in Indonesia: The Key Stakeholders' Perspectives*.

Sumarno, T., 2020. *Identifying and Modifying the Coal Bed Methane Fiscal Regime in Indonesia: A Comparative Study with the Australia and the U.S.. Doctorate*. University of Dundee.

Sunley, E., Baunsgaard, T. and Simard, D., 2002. Revenue from the Oil and Gas Sector: Issues and Country Experience. In: *IMF Conference*.

Tambunan, T., n.d. *PENGKAJIAN KEBIJAKAN INVESTASI RIIL DI INDONESIA*. [ebook] Jakarta: Pusat Studi Industri dan UKM, Universitas Trisakti. Available at: http://www.kadin-indonesia.or.id/enm/images/dokumen/KADIN-98-2929-16062008.pdf [Accessed 19 February 2020].

Thakur, P., 2017. *Advanced Reservoir and Production Engineering for Coal Bed Methane*. Cambridge, MA: Gulf Professional.

Thakur, P., Little, H. and Karis, W., 1996. Global coalbed methane recovery and use. *Energy Conversion and Management*, 37(6–8), pp. 789–794.

Tharakan, P., 2015. *Summary of Indonesia's Energy Sector Assessment*. No. 09, December 2015. Manila: Asian Development Bank, p. 14.

The Hon Josh Frydenberg MP, Treasurer of the Commonwealth of Australia, 2018. *Government Responds to Callaghan PRRT Review*. [online] Available at: https://ministers.treasury.gov.au/ministers/josh-frydenberg-2018/media-releases/government-responds-callaghan-prrt-review [Accessed 26 March 2020].

The Oxford Institute for Energy Studies, 2014. *The Future of Australian LNG Exports: Will Domestic Challenges Limit the Development of Future LNG Export Capacity?*. Oxford: The Oxford Institute for Energy Studies.

The Pembina Institute, 2003. *The Environmental Challenges of Coalbed Methane Development in Alberta*. [online] Drayton Valley: The Pembina Institute. Available at: https://www.pembina.org/reports/CBM_Final_April2006D.pdf [Accessed 11 February 2020].

The Wyoming Legislature, 2019. [online] Wyoleg.gov. Available at: https://www.wyoleg.gov/InterimCommittee/2015/SMTPresentationAG-DoR.pdf [Accessed 11 July 2019].

Theodori, G., 2009. *Paradoxical Perceptions of Problems Associated With Unconventional Natural Gas Development*. Texas: Southern Rural Sociological Association, pp. 97–117.

Thomas, M., 2013. *Unlocking Indonesia's CBM Potential | Exploration & Production*. [online] Epmag.com. Available at: http://www.epmag.com/unlocking-indonesias-cbm-potential-689451#p=full [Accessed 13 July 2016].

Thomson, S. and MacDonald, D., 2003. *The Application of Medium Radius Directional Drilling for Coal Bed Methane Extraction*. Researchgate.

Thurber, M. and Morse, R., 2015. *The Global Coal Market*. Cambridge: Cambridge University Press.

Tienhaara, K., 2006. Mineral investment and the regulation of the environment in developing countries: Lessons from Ghana. *International Environmental Agreements: Politics, Law and Economics*, 6(4), pp. 371–394.

Tole, L. and Koop, G., 2010. Do environmental regulations affect the location decisions of multinational gold mining firms?. *Journal of Economic Geography*, 11(1), pp. 151–177.

Tordo, S., 2007. *Fiscal Systems for Hydrocarbon, Design Issues*. 1st ed. [ebook] Washington: The World Bank. Available at: https://siteresources.worldbank.org/INTOGMC/Resources/fiscal_systems_for_hydrocarbons.pdf [Accessed 24 April 2016].

Treasury Laws Amendment (2019 Petroleum Resource Rent Tax Reforms No. 1) Act 2019. Royal Assented to 5 April 2019.

Treasury.gov.au. 2020. *Petroleum Resource Rent Tax: Review of Gas Transfer Pricing Arrangements | Treasury.Gov.Au*. [online] Available at: https://treasury.gov.au/consultation/c2019-t364690 [Accessed 1 April 2020].

Tully, J., 1996. *Coal Fields of the Conterminous United States*. U.S. Geological Survey Open-File Report OF, 96–92. United States: U.S. Geological Survey, pp. 92–96.

U.S. Department of Energy, 2002. *Powder River Basin Coalbed Methane Development and Produced Water Management Study*. Tulsa: U.S. Department of Energy.

U.S. Department of Energy, 2007. *Produced Water Management and Beneficial Use*. 2nd Semi-annual progress report. Colorado: Colorado Energy Research Institute.

U.S. Environmental Protection Agency (EPA), 2010. *Screener Survey Database (CBI)*. EPA-HQ-2008-0517 (DCN 07363).

United Nations Environment Program, 1997. *Environmental Management in Oil and Gas Exploration and Production*. [online] Paris: UNEP. Available at: https://www.google.com/url?sa=t&rct=j&q=&esrc=s&source=web&cd=18&cad=rja&uact=8&ved=2ahUKEwjz3dCk3cTnAhUUhlwKHdCwALkQFjARegQIChAB&url=https%3A%2F%2Fwedocs.unep.org%2Fbitstream%2Fhandle%2F20.500.11822%2F8275%2F-Environmental%2520Management%2520in%2520Oil%2520%2526%2520Gas%2520Exploration%2520%2526%2520Production-19972123.pdf%3Fsequence%3D2%2526isAllowed%3Dy&usg=AOvVaw3pbFYrR56jJRshfN98XN48 [Accessed 9 February 2020].

United Nations Fund for Population Activities, n.d. *The 2010 – 2035 Indonesian Population Projection*. [online] Singapore: UNFPA. Available at: https://indonesia.unfpa.org/sites/default/files/pub-pdf/Policy_brief_on_The_2010_–_2035_Indonesian_Population_Projection.pdf [Accessed 24 January 2020].

United Nations, 2017. *United Nations Handbook on Selected Issues for Taxation of the Extractive Industries by Developing Countries*. [online] New York: United Nations. Available at: https://www.google.com/url?sa=t&rct=j&q=&esrc=s&source=web&cd=1&cad=rja&uact=8&ved=2ahUKEwj9x7LDitHoAhVEr3EKHTqWCuMQFjAAegQIBRAB&url=https%3A%2F%2Fwww.un.org%2Fesa%2Fffd%2Fwp-content%2Fuploads%2F2018%2F05%2FExtractives-Handbook_2017.pdf&usg=AOvVaw3yELQwyKiQlkB11UPh1Vo9 [Accessed 23 March 2020].

United Nations, 2020a. [online] Unfccc.int. Available at: https://unfccc.int/process-and-meetings/the-paris-agreement/the-paris-agreement [Accessed 21 July 2020].

United Nations, 2020b. *All Ndcs*. [online] Www4.unfccc.int. Available at: https://www4.unfccc.int/sites/NDCStaging/Pages/All.aspx [Accessed 9 November 2020].

US Energy Information Administration, 2015. *International Energy Data and Analysis*. Indonesia. The US: US Energy Information Administration.

Utama, J. and Golubchenko, N., 2018. *Views on CBM Industry in Indonesia (Global Consultancy's Perspective).*

Vivoda, V., 2017. Determinants of Foreign Direct Investment in the Mining Industry. *Mining in the Asia-Pacific*, pp. 19–33.

Waeckerlin, E., 2005. The New Border War: CBM Development in the Powder River Basin of Wyoming and Montana. *Public Land & Resources Law Review*, 26, pp. 149–162.

Wajong, M., 2018a. *A Focus Group Discussion on CBM Industry in Indonesia.*

Wajong, M., 2018b. *Views on CBM Industry in Indonesia (Industry Association's Perspective).*

Weber, R., 1990. *Basic Content Analysis*. Newbury Park, Calif.: Sage.

William, J., Miligant, A. and Stubbs, T., n.d. *Coal Seam Gas Production: Challenges and Opportunities*. 1st ed. [ebook] Australia: Wentworth Group. Available at: http://wentworthgroup.org/wp-content/uploads/2013/12/BREE_Coal-seam-gas-production_WILLIAMS-etal-.pdf [Accessed 23 November 2016].

Williams, C., 2007. Research Methods. *Journal of Business & Economic Research*, 5(3).

Williams, R., Yurakov, E. and Ashelford, D., 2001. Gas Emissions Modelling of Gate Road Development. In: *Coal Operators' Geo-technology Colloquium.*

Willow, A., 2015. *Wells and Well-Being: Neoliberalism and Holistic Sustainability in the Shale Energy Debate*. [online] Taylor & Francis. Available at: http://www.tandfonline.com/doi/full/10.1080/13549839.2015.1017808 [Accessed 9 February 2020].

Wood, G. 2013. Connecting the dots: A systemic approach to evaluating potential constraints to renewable electricity technology deployment to 2020 and beyond in the United Kingdom' (PhD thesis, University of Dundee)

Geoffrey Wood, G. 2016. 'A systemic approach to renewable electricity technology deployment: The 'missing link' in optimising policy delivery in the UK?' in Raphael J. Heffron and Gavin F. M. Little (eds), *Delivering Energy Law and Policy in the EU and the US: A reader* (EUP 2016).

WoodMackenzie, 2015. *Sumatra CBM*. United Kingdom: Wood Mackenzie.

WoodMackenzie, n.d. *The Australia Fiscal Regimes.*

World Bank, 2020a. *Indonesia | Data*. [online] Data.worldbank.org. Available at: https://data.worldbank.org/country/ID [Accessed 11 November 2020].

World Bank, 2020b. *Overview*. [online] World Bank. Available at: https://www.worldbank.org/en/country/indonesia/overview [Accessed 11 November 2020].

World Bank. 2020c. *Overview*. [online] Available at: https://www.worldbank.org/en/country/indonesia/overview [Accessed 16 June 2020].

Worldometers.info. 2020. *Indonesia Population (2020) – Worldometer*. [online] Available at: https://www.worldometers.info/world-population/indonesia-population/ [Accessed 6 July 2020].

www.coagenergycouncil.gov.au. 2015. [online] Available at: http://www.coagenergycouncil.gov.au/publications/national-harmonised-regulatory-framework-natural-gas-coal-seams-annual-implementation [Accessed 9 July 2019].

Yahya, H., 2019. *Renewable Energy in the Era of Low Carbon Energy – Challenges & Opportunities -*.

Yan, T. and Curtin, R., 2010. The Relation Between Unit Nonresponse and Item Nonresponse: A Response Continuum Perspective. *International Journal of Public Opinion Research*, 22(4), pp. 535–551.

Yin, R., 2009. *Case Study Research Design and Methods*. 5th ed. United States of America: Sage Publications.

Index[1]

[1] Note: Page numbers followed by 'n' refer to notes.

T. B. Sumarno, *Just Energy Transitions and Coal Bed Methane*, Energy, Climate and the Environment, https://doi.org/10.1007/978-3-030-85490-4

Printed in the USA
CPSIA information can be obtained
at www.ICGtesting.com
LVHW051046030324
773397LV00005B/449

9 783030 854898